Electrotransport in Metals and Alloys

by J. N. Pratt and R. G. R. Sellors

About this Book

This book is part of the Diffusion and Defect Monograph Series edited by Y. Adda, A. D. Le Claire, L. M. Slifkin, and F. H. Wöhlbier. It has been written to serve as a first specialist text book in the area of electrotransport in solid and liquid metals and alloys. Monographs in this series are prepared when the essential content of the subject is well established and there is agreement on its scope, its limitations and the likely general direction its future development may take. Thus these books are of quite lasting value as they deal with subjects that have reached a certain stage of maturity.

Following a detailed discussion of the theoretical background, the authors present the various phenomenological and theoretical models of electrotransport in solid and liquid metallic systems. The available experimental techniques are described and all experimental data are summarized. Discussion of the experimental results leads to a critical analysis of the theoretical models and a presentation of current basic problems. The book furthermore covers some of the aspects involved in the technological utilization of electrotransport phenomena.

Electrotransport in metals and alloys is of much technical interest in the area of metal purification and in understanding certain causes of failure in current carrying devices; in addition, study of electrotransport behavior yields fundamental information concerning ionic states and transport processes in materials. The subject therefore is of importance to material scientists and solid state physicists, as well as to engineers working with electrical currents and metallurgists engaged in metal purification work.

TABLE OF CONTENTS

1. Introduction

The application of an electric field to a metal, alloy or other material may result in the transport of matter as well as the flow of electrons. This phenomenon, variously described as electrotransport, electromigration or electrodiffusion, is most commonly manifested in the separation of the components of an alloy which results from their different induced rates or directions of migration. In pure elements, an applied field may similarly produce a directed displacement of vacancies or interstitials (self-transport) or of different isotopes (the Haeffner effect).

Electrotransport was first reported by Gerardin[1] in 1861. This and all the subsequent early studies were concerned with liquid alloys and it was not until the 1930's that the much slower process of electrotransport in the solid state was demonstrated and investigated. Since that time, and particularly in recent years, it has received a considerable amount of experimental and theoretical attention. Interest in the phenomenon has been stimulated by its possible significance as a materials purification technique[2,3,4,5,109], as a cause of deleterious effects in current carrying devices[6,7,8,115], or as a source of fundamental information concerning ionic states and transport processes in materials. Unfortunately, many of the data are at best only poorly quantitative. As a consequence of this and of the complexity of the problems involved, fundamental and even empirical understanding of the phenomenon remains imperfect and incomplete.

The numerous publications on this subject are widely dispersed through the literature. Electrotransport work has been reviewed previously by Schwarz[9], Jost[10], Seith[11], Heumann[12], Verhoeven[13], Belashchenko[14] and Huntington[86], most of whom have made selected collections of data available at the time. During a recent research programme the present comprehensive bibliography and collection of numerical data was prepared for convenient reference. Information from previous reviews has been combined and augmented by the addition of results taken from the many more-recently published papers.

2. Theoretical Background

2.1 General Aspects

Phenomenologically, an ion migrating under the influence of an applied field may be considered as acted upon by two forces:

(i) the "field force" or "electrostatic force" due to the direct interaction of the field with the charged ion; this force will be cathode-directed for a positive ion and anode-directed in the case of one of negative charge.

(ii) the "drag", "friction force" or "electron wind" due to the momentum exchange between the current carriers (electrons or positive "holes") and the ions. This force will be anode-directed when the current is predominantly carried by electrons, but positive "holes" may give rise to cathode-directed momentum exchange[15,52].

In the case of a simple metal consisting of positive ions and an electron "gas", this may be represented schematically by

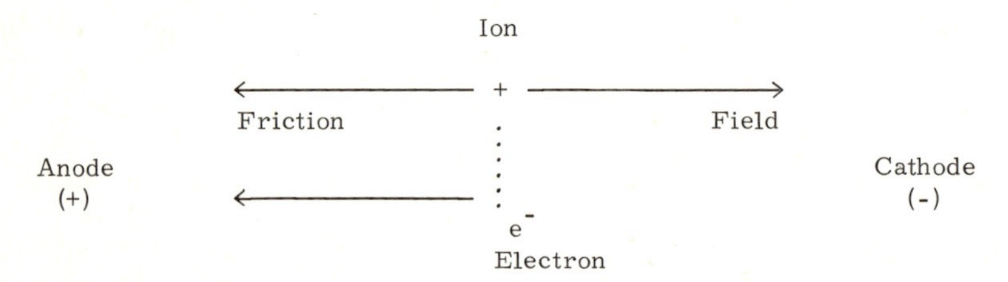

If the resultant force on the ion is represented by F_i, then

$$F_i = \text{Field force} - \text{Drag (friction) force}. \quad (1)$$

Assuming that the friction force, F_{ei}, is directly proportional to the electric field, E, then one may write

$$F_{ei} = \delta_{ei} E , \tag{2}$$

where δ_{ei} represents a friction coefficient. The resultant net force on an ion of actual valency Z_i may then be represented as

$$F_i = eE(Z_i - \frac{\delta_{ei}}{e}) \tag{3}$$

or

$$F_i = eEZ_i^o , \tag{4}$$

where $Z_i^o = (Z_i - \frac{\delta_{ei}}{e})$ may be regarded as the effective valency of the component.

Now the velocity of transport of an ion, V_i, may be written as

$$V_i = B_i F_i , \tag{5}$$

where B_i is the absolute mobility (terminal velocity/unit force) and

$$B_i = \frac{D_i}{fkT} , \tag{6}$$

where D_i is the diffusion coefficient and f the lattice correlation coefficient allowing for non-random diffusion jumps in a crystalline solid.

Thus from equations (3), (5) and (6)

$$V_i = \frac{D_i}{fkT} eE(Z_i \frac{\delta_{ei}}{e}) , \tag{7}$$

or defining an "electric mobility", U_i, as the transport velocity per unit field

$$U_i = \frac{D_i e}{fkT} (Z_i - \frac{\delta_{ei}}{e}) = \frac{D_i e}{fkT} Z_i^o . \tag{8}$$

Clearly the sign and magnitude of the effective valency or the rate and direction of migration are determined by the directions and relative predominance of the field and drag forces. Since, in principle, D_i may be obtained from diffusion studies and U_i from electrotransport measurements, Z_i^o the effective valency may be obtainable experimentally; the true charge Z_i can only be determined from this if δ_{ei} can be evaluated theoretically or semi-empirically. Although employed in the analysis of some early work, mobility equations of the above form strictly apply only to pure materials. Lattice correlation effects are involved in a more complex fashion in the case of migration in solid substitutional alloys[45,87].

Electrotransport data may alternatively be conveniently represented in terms of a transport number, t_i, where

$$t_i = \frac{U_i C_i F^*}{\sigma} \tag{9}$$

and σ = conductivity, F^* = Faraday's constant and C_i = moles of i per unit volume. Some Russian workers have reported results in the form of an "electro-diffusion coefficient", K given by the relation

$$\ln \frac{C_x}{C_o} = -K \frac{\Delta \Phi_x}{T}, \tag{10}$$

where C_x and C_o are the steady state concentrations at co-ordinates O and x and $\Delta \Phi_x$ the potential drop. The basis for this treatment can be appreciated from equation (31), Section 2.2.2.

Except in the cases of interstitial solutes, which may migrate rapidly, and very dilute solutions, where the mobility of the solvent may be regarded as effectively zero, the simple observation of field induced concentration differences does not yield individual transport numbers and even the direction of migration remains indeterminate. Thus, for the more general cases, electrotransport data are frequently limited to a knowledge of the differential electric mobilities, U_{12}, between the components 1 and 2 of an alloy, where

$$U_{12} = \frac{V_1 - V_2}{E} = U_1 - U_2 . \tag{11}$$

Such data must be combined with measurements of total mass transport by means of weighing or inert or radioactive marker techniques in order to obtain individual true transport numbers.

There are essentially two types of electrotransport experiments. In those performed in the absence of a concentration gradient, the measured velocity of a component is solely due to the net force arising from the field and drag forces and data from these experiments are frequently reported as transport numbers, which may be related to electrical mobilities by equation (9). The second type of experiment is continued until a steady state is reached when the above net force is counterbalanced by the field-induced chemical potential (concentration) gradient; in these cases results are most frequently presented as effective valencies which, in turn, may be related to electrical mobilities by equation (8).

In the past many data have been interpreted on the assumption that the electric field force was predominant over the electron drag force. In some cases this leads to the implication of improbably large actual valencies being possessed by the component ions. It has, therefore, become increasingly apparent that such field force interpretations are unlikely to be valid and that the electron drag force is probably the major factor affecting electrotransport. Although the existence of this force was recognised long ago by Skaupy[16] (1914) and by Lewis et al[17] (1915), the first quantitative treatments, by Wagner[18] and by Schwarz[19], neglected this and attempted essentially classical treatments of the problem. The last investigator did, however, recognise the complexity of the problem by introducing the idea of "electrohydrostatic forces" acting on the components in addition to the classical field force. A more recent attempt to treat the problem classically is represented by Drakin's[20] application of classical thermodynamics to electrotransport.

2.2 Phenomenological Models

2.2.1 Baranowski's treatment of dilute liquid alloys

Most authors have recognised that electromigration is concerned with steady-state rather than equilibrium conditions and have therefore employed the methods of irreversible thermodynamics[21] in attempting to establish valid phenomenological models. Typical of these is Baranowski's[22-28] treatment of dilute liquid alloys which makes use of the Onsager linear flux force relationships.

Considering, for simplicity, a binary system, the fluxes of components and electrons may be expressed by the equations

$$J_k = \sum_{i=1}^{3} L_{ki} X_i \quad (k = 1,\ 2,\ 3), \tag{12}$$

where the indices 1 and 2 denote the component ions and 3 denotes electrons; L_{ki} are the Onsager coefficients, obeying the usual reciprocal relationship $L_{ki} = L_{ik}$; X_i represents the thermodynamic forces which may in turn be written as

$$X_i = -\operatorname{grad} \mu_i - Z_i \operatorname{grad} \Phi, \tag{13}$$

where μ_i is the molar chemical potential of the component i and Φ is the electrical potential.

Assuming electroneutrality and incompressibility, so that

$$\sum_{i=1,2} Z_i C_i - C_3 = 0 \qquad (C_i = \text{concentration}) \tag{14}$$

and

$$\sum_{i=1}^{3} J_i \overline{V}_i = 0 \qquad (\overline{V}_i = \text{partial volume}) \tag{15}$$

and defining e_i^*, transfer parameters representing the interaction of the flow of electrons with that of the component ions in the absence of an electric field, by

$$e_1^* = \left(\frac{J_3}{J_1}\right)_{X_2 = X_3 = 0} = \frac{L_{13}}{L_{11}} \tag{16}$$

and

$$e_2^* = \left(\frac{J_3}{J_2}\right)_{X_2 = X_3 = 0} = \frac{L_{23}}{L_{22}}, \tag{17}$$

leads to expressions for the ion fluxes in a binary alloy which are of the form

$$J_1 = L_{11}(X_1 - \frac{\bar{V}_1}{\bar{V}_2} X_2 + e_1^* X_3) \tag{18}$$

and for the electron flux

$$J_3 = L_{11} e_1^* (X_1 - \frac{\bar{V}_1}{\bar{V}_2} X_2) + L_{33} X_3 . \tag{19}$$

Assuming grad $\mu_e = 0$ and considering a dilute solution, where, if 1 denotes the solute, grad $\mu_2 \simeq 0$, then equations (13) become

$$X_1 = - \text{grad } \mu_1 - Z_1 \text{ grad } \Phi , \tag{20}$$

$$X_2 = - Z_2 \text{ grad } \Phi \tag{21}$$

and $$X_3 = F^* \text{ grad } \Phi \tag{22}$$

Substituting in equation (18) for the solute flux gives

$$J_1 = L_{11} (- \text{grad } \mu_1 - Z_1 \text{ grad } \Phi + \frac{\bar{V}_1}{\bar{V}_2} Z_2 \text{grad } \Phi + e_1^* F^* \text{ grad } \Phi). \tag{23}$$

Hence, by considering the steady state ($J_1 = J_2 = 0$), is obtained the relation

$$\frac{\text{grad } \mu_1}{\text{grad } \Phi} = \frac{\bar{V}_1}{\bar{V}_2} Z_2 - Z_1 + e_1^* F^* \tag{24}$$

At the steady state, the induced gradient of chemical potential is exactly balanced by the net force due to the applied field, i.e. grad $\mu_1 = Z_1^o e.$ grad Φ. It is clear, therefore, that the R.H.S. of equation (24) again has the significance of effective charge (per g. ion) of component 1. The final term $e_1^* F^*$ represents the effective charge with respect to the electron wind, since e_1^* is dependent on L_{13}, the Onsager coefficient representing electron-ion interaction; the remaining terms represent the excess charge of the solute ions giving rise to the direct field force.

2.2.2 Belashchenko and Zhukovitskii's treatment

A related treatment of a two-component system has been given by

Belashchenko and Zhukovitskii[29]. For a one-dimensional system, the forces X_i of the Onsager equations (13) are expressed by

$$X_i = -\frac{\delta \mu_i}{\delta x} - Z_i^o e \frac{\delta \Phi}{\delta x} , \tag{25}$$

where μ_i represents the chemical potential of the ions or electrons.

The electroneutrality condition is represented by

$$\sum Z_i J_i = J , \tag{26}$$

where J is the external current density flowing through the system. Considering the particular case where $\bar{V}_1 = \bar{V}_2$, so that $J_1 + J_2 = 0$, it is shown that

$$\begin{aligned} L_{11} &= - L_{12} = L_{22} \\ L_{13} &= - L_{23} \quad . \end{aligned} \tag{27}$$

The chemical potentials of the component atoms are introduced through the relations

$$\mu^o_{atom} = \mu_{ion} - Z_{ion} \mu_{electron} \tag{28}$$

and the Onsager equations are then solved taking account of conditions (26), (27) and (28) and the Gibbs-Duhem relation between the thermodynamic properties of the components. This yields the result

$$J_1 = N_1 U_1 E - D_1 \frac{dN_1}{dx} , \tag{29}$$

where N_1 here indicates mole fraction of component 1. Belashchenko and Zhukhovitskii define the effective charge on the ion by the relation

$$U_i = \frac{D_i}{kT} \cdot Z_i^o \; e \frac{\delta \ln N_i}{\delta \ln a_i} , \tag{30}$$

where a_i is the thermodynamic activity of the ith component. Consideration of the steady state conditions, so that $J_1 = 0$, yields for each component

$$\frac{d. \ln a_i}{dx} = \frac{Z_i^o eE}{kT} , \tag{31}$$

which is equivalent to the Baranowski equation (24). The treatment has been generalized[30] to apply to an n-component system.

2.2.3 The Adda and Philibert treatment of solids

A comprehensive phenomenological treatment of electrotransport in ionic crystals, pure metals and interstitial and substitutional solutions has been presented by Adda and Philibert[87]. The general equation for the rate of creation of entropy per unit volume (θ') is re-expressed in the form

$$\theta' = J_q' \cdot \mathrm{grad}(\frac{1}{T}) - \frac{1}{T}\sum_i J_i \cdot \mathrm{grad}\,\tilde{\mu}_i - \frac{1}{T} J_e \cdot \mathrm{grad}\,\tilde{\mu}_e - \frac{\theta'_V \tilde{\mu}_V}{T} \qquad (32)$$

where J_q' represents the thermal flux, J_i the flux of component i, J_e the electron flux and θ'_V the rate of creation of vacancies. The terms $\tilde{\mu}$ denote the respective thermodynamic potentials in the presence of the applied field and may be written

$$\tilde{\mu}_i = \mu_i + Z_i e \Phi ,$$

or for an electron

$$\tilde{\mu}_e = E_{Fermi} - e\Phi ,$$

while for a non-charged particle such as a vacancy

$$\tilde{\mu}_V = \mu_V .$$

Following the theorem of Prigogine that the rate of creation of entropy is constant, the relations between the fluxes and forces in the various cases are simply deduced from equation (32).

For an interstitial solute B in a metallic solvent, since the displacement of the solvent atoms may be neglected to a first approximation, one may thus write

$$J_B = - \frac{L_{BB}}{T} \mathrm{grad}\,\tilde{\mu}_B - \frac{L_{Be}}{T} \mathrm{grad}\,\tilde{\mu}_e . \qquad (33)$$

That is to say,

$$J_B = -\frac{kL_{BB}}{n_B}\left(1 + \frac{\partial \log\gamma_B}{\partial \log N_B}\right) \text{grad}\, n_B - \frac{L_{BB}}{T} Z_B e \,.\, \text{grad}\, \Phi$$

$$- \frac{L_{Be}}{T} \,.\, \text{grad}\, E_{Fermi} + \frac{L_{Be}}{T} e \,.\, \text{grad}\, \Phi, \tag{34}$$

where γ_B = activity coefficient of B.

Since for a metallic solvent grad $E_{Fermi} = 0$, equation (33) may be written as

$$J_B = - D_B^A \,\text{grad}\, n_B - \frac{1}{T}\left[L_{BB} Z_B - L_{Be}\right] e \,.\, \text{grad}\, \Phi, \tag{35}$$

where the diffusion coefficient of B in A, $D_B^A = \frac{kL_{BB}}{n_B}\left(1 + \frac{\partial \log\gamma_B}{\partial \log N_B}\right)$.

From inspection of equation (35) it is clear that the second term represents the effect of the field, so that the electrical mobility is seen to be equal to

$$u_B = \frac{L_{BB}}{n_B T}\left[Z_B - \frac{L_{Be}}{L_{BB}}\right] e\,. \tag{36}$$

But if the solution of B is sufficiently dilute to obey Henry's law, then γ_B is constant and $D_B^A = kL_{BB}/n_B$. Equation (36) then becomes

$$u_B = \frac{D_B^A}{kT}\left[Z_B - \frac{L_{Be}}{L_{BB}}\right] e$$

$$= \frac{D_B^A}{kT} \,.\, Z_B^o \,.\, e\,, \tag{37}$$

thus giving the Nernst-Einstein relation written with the effective charge

$$Z_B^o = Z_B - \frac{L_{Be}}{L_{BB}}\,.$$

For a substitutional alloy, A - B, where diffusion occurs by a vacancy mechanism, the flux of the components, measured with respect to the lattice reference, may be expressed in the form

$$J_A = -\frac{L_{AA}}{T}\,\mathrm{grad}(\tilde{\mu}_A - \tilde{\mu}_V) - \frac{L_{AB}}{T}\,\mathrm{grad}(\tilde{\mu}_B - \tilde{\mu}_V) - \frac{L_{Ae}}{T}\,\mathrm{grad}\,\tilde{\mu}_e$$

$$J_B = -\frac{L_{BA}}{T}\,\mathrm{grad}(\tilde{\mu}_A - \tilde{\mu}_V) - \frac{L_{BB}}{T}\,\mathrm{grad}(\tilde{\mu}_B - \tilde{\mu}_V) - \frac{L_{Be}}{T}\,\mathrm{grad}\,\tilde{\mu}_e$$

$$J_V = -(J_A + J_B)\,. \tag{38}$$

Considering, for example, component B and introducing the potentials $\tilde{\mu}$ defined as above, this yields

$$J_B = -\frac{L_{BA}}{T}(\mathrm{grad}\,\mu_A - \mathrm{grad}\,\mu_V) - \frac{L_{BB}}{T}(\mathrm{grad}\,\mu_B - \mathrm{grad}\,\mu_V) - \frac{L_{Be}}{T}\,\mathrm{grad}\,E_{Fermi} - \frac{1}{T}(L_{BA}Z_A + L_{BB}Z_B - L_{Be})\,e\cdot\mathrm{grad}\,\Phi\,. \tag{39}$$

But for a dilute solution of B, $\mu_B = kT\log\gamma_B$ and $\mathrm{grad}\,\mu_A = 0$; as before $\mathrm{grad}\,E_{Fermi} = 0$ for the metallic state and assuming thermal equilibrium for the vacancies, $\mathrm{grad}\,\mu_V = 0$. Thus we have, under these conditions:

$$J_B = -\frac{kL_{BB}}{n_B T}\cdot\mathrm{grad}\,n_B - \frac{1}{T}\left[L_{BA}Z_A + L_{BB}Z_B - L_{Be}\right]e\cdot\mathrm{grad}\,\Phi\,. \tag{40}$$

Following Baranowski[88], the coefficient L_{Be} originating from the interaction of the B atoms with the charge carriers may be defined by the relation

$$L_{Be} = L_{BB}Z_B^* + L_{BA}Z_A^*\,, \tag{41}$$

where Z_B^* and Z_A^* represent the valencies relative to the <u>friction</u> force (i.e. they are equivalent to δ_{eB}/e and δ_{eA}/e where δ is the friction coefficient of equation (2)). Recalling also that $(kL_{BB}/n_B) = D_B^A$, the hetero-diffusion coefficient of B in A, equation (41) becomes

$$J_B = -D_B^A\,\mathrm{grad}\,n_B + \frac{n_B\cdot D_B^A}{kT}\left[(Z_B - Z_B^*) + \frac{L_{BA}}{L_{BB}}(Z_A - Z_A^*)\right]e\cdot\mathrm{grad}\,\Phi\,. \tag{42}$$

From this it follows that the electric mobility is given by

$$u_B = \frac{D_B^A}{kT}\left[Z_B^o + \frac{L_{BA}}{L_{BB}}Z_A^o\right]e \tag{43}$$

This expression may be compared with the Nernst-Einstein relationship which would yield

$$u_B = \frac{D_B^A}{kT} (Z_B^o) e . \tag{44}$$

This comparison shows that the latter relation is only valid when the fluxes are independent since the correction term involves the cross-coefficient L_{BA} which describes the coupling between the fluxes of A and B atoms.

Equation (43) may be expressed in the Nernst-Einstein form as

$$u_B = \frac{D_B^A}{kT} Z_B^{oo} e \tag{45}$$

by defining an "apparent effective valency" $Z_B^{oo} = Z_B^o \left[1 + \frac{L_{BA}}{L_{BB}} \frac{Z_A^o}{Z_B^o} \right]$.

Clearly experimental measurements on substitutional solutions will strictly yield apparent effective valencies (Z^{oo}) rather than the true effective valencies (Z^o) required for comparison with the theoretical models described below. The significance of the correction has been examined by Van Doan and Brebec[84,85] (see Section 2.5). Provided L_{BA}/L_{BB} is small then $Z_B^{oo} = Z_B^o$.

For the case of <u>self-transport</u> one may write $B = A^*$ so that equation (43) becomes

$$u_A = \frac{D_A^A}{kT} Z_A^o \left(1 + \frac{L_{AA^*}}{L_{A^*A^*}}\right) e \tag{46}$$

But $\left(1 + \frac{L_{AA^*}}{L_{AA}}\right) = \frac{1}{f_A}$ [87], where f_A is the correlation factor, therefore

$$u_A = \frac{D_A^A}{f_A kT} \cdot Z_A^o e = \frac{D_A^A}{kT} \cdot Z_A^{oo} e \tag{47}$$

where $Z_A^{oo} = Z_A^o / f_A$.

2.2.4 Klemm's treatment of isotope separation

The particular case of the separation of isotopes in an electric field has been considered by Klemm[31,32]. Denoting a "coefficient of friction" between the ith and k^{th} types of particle by r_{ik}, the relative velocities between the same particles by v_{ik}, and the degree of ionization of a component by ξ_i, he writes from the condition of equality of electric and "friction" forces for each type of particle:

$$- \xi_1 Z_1 F^* \text{ grad } \Phi = r_{12} N_2 V_{12} + r_{1n} N_n V_{1n} \quad , \tag{48}$$

$$- \xi_2 Z_2 F^* \text{ grad } \Phi = r_{21} N_1 V_{21} + r_{2n} N_n V_{2n} \tag{49}$$

and

$$\xi_n Z_n F^* \text{ grad } \Phi = r_{n1} N_1 V_{n1} + r_{n2} N_2 V_{n2} \quad , \tag{50}$$

where the subscripts 1, 2 and n indicate respectively the light and heavy isotopes and the electrons; as before Z_i is the actual valency and N_i the mole fraction.

By assuming that $r_{ik} = r_{ki}$ and $Z_1 = Z_2 = Z$, and defining the relationships

$$N_1 + N_2 = N_p, \; N_p + N_n = 1, \; N_n = ZN_p \tag{51}$$

and

$$\xi_p = (N_1 \xi_1 + N_2 \xi_2)/N_p,$$

Klemm solves equations (48)-(50) to obtain

$$\frac{V_{12}}{V_{pn}} = \left(\frac{\xi_1 - \xi_2}{p} - \frac{r_{1n} - r_{2n}}{r_{pn}}\right) \Big/ \left(1 + \frac{r_{12}}{Zr_{pn}}\right) \quad . \tag{52}$$

Introduction of estimated values for the terms involved, however, results in discrepancies in sign and magnitude between the predicted isotope migration and experimental observation. Klemm therefore postulates that the ions of each isotope exist in two states - mobile (activated) and immobile. Equations (48)-(50) are thus replaced by four equations describing respectively mobile ions of the light isotope, mobile ions of the heavy isotope, immobile ions and electrons. Using the subscript b to denote mobile ions and u for immobile, he then obtains the relation:

$$\frac{V_{12}}{V_{pn}} = \frac{N_b}{N_p} \cdot \frac{V_{bu}}{V_{pn}} \cdot \frac{r_{2bu} - r_{1bu}}{r_{bu}} \Big/ \left(1 + \frac{N_b}{N_u} \frac{r_{1b2b}}{r_{bu}}\right) . \tag{53}$$

The phenomen of self-transport may also be treated by considering a pure metal to consist of mobile and immobile species and by assuming that r_{bn} and r_{un} are negligible relative to r_{bu}. In this way, Klemm has derived the equation:

$$\frac{V_{bu}}{V_{pn}} = \frac{(\xi_b r_{un} - \xi_u r_{bn})}{(N_b \xi_b + N_u \xi_u)} \cdot \frac{N_p}{r_{pn}} \cdot \frac{Zr_{pn}}{r_{pu}} \qquad (54)$$

This equation has been re-arranged by Verhoeven[13] for comparison with other treatments. The average electron velocity may be related to the conductivity by $V_{pn} = \sigma E/n_e e$, the friction coefficient to the self-diffusion coefficient by $r_{bn} = RT/D$ and the degree of ionization ξ_i converted to effective valence Z_i; n_e is the number of electrons per unit volume. Taking note of the balance of charge in the system and regarding the immobile ions as stationary, the velocity of self transport is then given by

$$V_{bu} = \frac{\sigma E}{F^* n_e} \left(\frac{D}{KT}\right) r_{un} \left(Z_b - Z_u \frac{r_{bn}}{r_{un}}\right) , \qquad (55)$$

which, combined with the uncorrelated form of equation (6), yields for the net force on the ion

$$F = eEK\left(Z_b - Z_u \frac{r_{bn}}{r_{un}}\right) , \qquad (56)$$

where $$K = \frac{\sigma r_{un}}{F n_e e} = 1^{13} .$$

The equation is also applicable to the electromigration of interstitials by considering these as the activated ion.

2.3 Semi-empirical Models

2.3.1 The Mangelsdorf treatment of liquid alloys

Various attempts have been made to examine in detail the forces acting on electromigrating ions and to establish the exact significance of effective charges. One quasi-phenomenological treatment is that of Mangelsdorf[33] whose model treats the ions in a classical manner and the electrons as a quantum fluid. He expresses the average electrical force on a particular ion species (s) by

$$F_s = (z_s + b_s + a_s)eE^o + \sum_k (d_{sk} + c_{sk}) J_{ek}. \tag{57}$$

Here $z_s e$ is the net charge on the ion (core electrons plus nucleus) treated as an inseparable but polarizable unit and $z_s eE^o$ the normal field force due to the direct action of the applied field (E^o) on the ion; $b_s eE^o$ represents a possible force due to the polarization of the collective electrons by the field and $a_s eE^o$ a force due to the polarization of all the other ions by the field; $d_{sk} J_{ek}$ is a force due to the current (J_e) polarization of the collective electrons and $C_{sk} J_{ek}$ that due to the current polarization of the other ions.

The total electrical force on all ions taken together then becomes

$$F_{\text{all ions}} = \sum_s n_s (z_s + b_s)eE^o + \sum_{s,k} n_s d_{sk} J_{ek} , \tag{58}$$

where all the ion-ion interactions, represented by the coefficients a_s and c_s in equation (57), have disappeared by mutual cancellation.

Since for a metal, being electrically neutral, the total charge of the electron system is just equal and opposite to the total charge of the ions it follows that

$$F_{\text{all ions}} = - F_{\text{electrons}} .$$

But with a constant field E^o the electron current will achieve a steady state value (J_e^o) and $F_{\text{electron}} = 0$ since there is then no net acceleration of the collective electrons. Thus one may write

$$\sum_s n_s(z_s + b_s)eE^o + \sum_{s,k} n_s d_{sk} J_{ek}^o = 0 . \tag{59}$$

Since for a metal the ionic current is negligible and the electronic current essentially constitutes the total current, then

$$- eJ_e^o \sim I^o = \sigma E^o , \qquad (60)$$

where σ is the conductivity, so that (59) may be rewritten to give

$$\sigma = e^2 \sum_s n_s (z_s + b_s) / \sum_{s,k} n_s d_{sk} \frac{J^o_{ek}}{J^o_e} . \qquad (61)$$

Comparison with the familiar basic expression for conductivity, $\sigma = e^2 n_{eff} \tau/m$, reveals that $\sum_s n_s (z_s + b_s)$ corresponds to n_{eff}, the effective number of free electrons, where $\sum_s n_s b_s$ represents the forces on the ions due to the polarization of the collective electrons by the field; $\sum_s n_s d_{sk} \frac{J^o_{ek}}{Je}$ corresponds to $^m/\tau$, the "frictional coefficient" for electron motion. The relation of the treatment to earlier phenomenological equations (e.g. (3), (56)), is demonstrated by defining the terms

$$Z_s = z_s + b_s + a_s$$

and
$$R_s = \sum_\kappa (d_{sk} + c_{sk}) \frac{J^o_{ek}}{J^o_e}$$

whence, since $\sum_s n_s a_s = \sum_{s,k} n_s c_{sk} \cdot \frac{J^o_{ek}}{J^o_e} = 0$ (ion interaction terms cancelling), in the steady state, equation (61) becomes

$$\sigma = e^2 \langle Z \rangle / \langle R \rangle \qquad (62)$$

and (57) becomes

$$F_s = eE^o (Z - R_s \frac{\langle Z \rangle}{\langle R \rangle}) \qquad (63)$$
$$= eE^o \text{"}Z_s\text{"} \quad ;$$

$\langle Z \rangle$ and $\langle R \rangle$ denote average values.

The coefficients Z_s and R_s may vary with composition since a solute ion may be expected to interact directly with immediately neighbouring solvent or solute ions and indirectly with all ions within an electron mean free path distance. Considering the direct ("chemical") interactions the more important, Mangelsdorf visualizes in effect the migration of clusters, since the electrical forces are exerted on the solute plus the surrounding solvent ions with which the solute is directly interacting. The addition of a solute (A) may be regarded as inducing a number of solvent ions, $n_B(Z, R)$, which fall within the solute's range of influence, to assume values of Z and R different from the pure solvent values Z_B^o and R_B^o; the modified value of Z_B is expressed as

$$Z_B = 1/N_B \left[N_B Z_B^o + N_A \int (Z - Z_B^o) n_B(Z, R).\, dR.\, dZ \right], \qquad (64)$$

with a similar expression for R_B.

At infinite dilution, $N_A = 0$, $Z_B = Z_B^o$ and this yields

$$(dZ_B/dN_A)_{N_A=0} = \int (Z - Z_B^o) n_{B_{(N_A=0)}}.\, dR.\, dZ \ , \qquad (65)$$

with a similar result for R_B.

Combining (65) with the logarithmic derivative of (62) with respect to N_A, and setting $N_A = 0$, so that $\langle Z \rangle = \langle Z_B^o \rangle$ and $\langle R \rangle = R_B^o$, leads to the relation, for dilute solutions:

$$\left[(1/\sigma_o)(d\sigma/dN_A) \right]_{N_A=0} = (1/Z_B^o)(\text{"}Z_A\text{"} + \int \text{"}Z\text{"} n_{B(N_A=0)}.\, dR.\, dZ) \ , \qquad (66)$$

where $\text{"}Z\text{"} = Z - R\,(Z_B^o/R_B^o)$.

Bringing in equation (63) yields, for the total electric force on the migrating solute and associated neighbours,

$$F_{total} = Z_B^o e E^o (1/\sigma)(d\sigma/dN_A)_{N_A} = 0 \ . \qquad (67)$$

But this force can also be equated to $Z_A^o e E^o$ where Z_A^o represents the "effective valency" of the migrating solute as may be determined from appropriate migration velocity and diffusion coefficient measurements. It will be seen that a plot of such Z_A^o values, for different solutes in a given solvent, versus $(1/\sigma)(d\sigma/dN_A)_{N_A=0}$ for the same alloys should produce a straight line of slope Z_B^o, the mean charge of the solvent ions.

2.3.2 de Gennes's treatment of isotope separation

A more theoretical treatment of isotope separation has been attempted by de Gennes[34] whose analysis of electron-ion interactions in a molten metal containing two isotopes indicates that the effective cross-section of the light isotope is greater than that of the heavy; in the presence of a field the first is thus entrained by the electrons. The quantitative treatment assumes a knowledge of the number of free electrons, of the relation between the coefficients of diffusion and viscosity, and of the mean frequency of oscillation of an ion, but a knowledge of the electronic mass is not involved. Under conditions of steady state transport, de Gennes considers the net transporting force on an ion, due to the direct-field and electron-wind effects, to be balanced by a "viscous" force arising from the normal interactions with surrounding ions; the latter force is expressed as the product of the mean momentum of the ion ($M_i\bar{v}_i$) and the viscosity of the metal for the isotope i, which may be defined as η_i. One thus has the relation

$$M_i\bar{v}_i\,\eta_i = ZeE - F_{ei} \quad . \tag{68}$$

Assuming the activation energy for the two processes to be the same, the viscosity and the coefficient of self diffusion (D) are related by

$$\frac{1}{\eta} = C\frac{MD}{kT} \quad , \tag{69}$$

where η and M represent mean values of η_i and M_i and C is a numerical constant which can be demonstrated to be approximately unity. Combining (68) and (69) thus yields

$$\bar{v}_i\left(\frac{kT}{D}\right)\frac{1}{C} = ZeE - F_{ei} \quad . \tag{70}$$

The electron wind force, F_{ei}, is estimated assuming the existence of well-defined mean free times for the electrons. Thus the total electron collisions per ion per unit time may be equated to the sum of collisions on the separate isotopes, as

$$\frac{Z}{\tau_e} = \frac{N_1}{\tau_1} + \frac{N_2}{\tau_2} \quad , \tag{71}$$

where N_1 and N_2 are the mole fractions of the isotopes, τ_e is the electron mean free time and τ_1 and τ_2 are the mean times between electron collisions on species 1 and 2. But

$$\frac{N_1}{\tau_1} + \frac{N_2}{\tau_2} = N_1(\frac{1}{\tau_1}) + N_2(\frac{1}{\tau_2})$$

$$= \text{weighted average collision rate}$$

$$= (\text{say}) \frac{1}{\tau} \quad ,$$

where τ is average time between impacts on ions.

Thus
$$\frac{Z}{\tau_e} = \frac{1}{\tau} \quad . \tag{72}$$

If at each collision an electron transfers to the ion only that part of its momentum acquired since its previous collision i.e. $eE\tau_e$, then

$$F_{ei} = \frac{eE\tau_e}{\tau_i} \tag{73}$$

or, from (73) and (72),

$$F_{ei} = eEZ\frac{\tau}{\tau_i} \quad .$$

Substituting in equation (70) then yields for the mean velocity of the species:

$$\bar{v}_i = C(\frac{D}{kT})\, eEZ(\frac{\tau_i - \tau}{\tau_i}) \tag{74}$$

or, for the differential velocity,

$$\Delta v = \bar{v}_1 - \bar{v}_2 = C(\frac{D}{kT})eEZ\tau \, . \, \Delta(\frac{1}{\tau}) \quad , \tag{75}$$

where $\Delta(\frac{1}{\tau}) = (\frac{1}{\tau_2} - \frac{1}{\tau_1})$ and it is assumed that $\tau^2 = \tau_1 \tau_2$.

By considering the absorption and emission of energy by ions oscillating with a frequency ω within a cage of neighbours, but without correlation between them, it is shown that the relative variation of $^1/T_i$ between the two isotopes is

$$\Delta(\frac{1}{T})/\frac{1}{T} = \frac{\Delta\omega}{\omega}\frac{1}{6}(\frac{\hbar\omega}{kT})^2$$

$$= \frac{1}{12}\frac{\Delta M}{M}(\frac{\hbar\omega}{kT})^2$$

or
$$= \frac{1}{12}\frac{\Delta M}{M}(\frac{\Theta}{T})^2 \qquad (76)$$

where Θ is the Debye temperature.

Using (75) and (76) then finally yields, for the differential mobility,

$$U_{12} = \frac{\Delta v}{E} = \frac{C}{12}(\frac{D}{kT})eZ(\frac{\Theta}{T})^2\frac{\Delta M}{M} . \qquad (77)$$

Physically the treatment indicates that the lighter isotope is slightly more "de-localized" and, having a greater amplitude of vibration, perturbs the movement and feels the impact of electrons more and hence is entrained by them. Theoretical values, calculated using this model, show poor quantitative agreement with experiment; thus for liquid gallium, for example, it suggests effects weaker, by a factor of two, than observed. Qualitatively, however, it predicts accumulation of the light isotope at the anode in accord with all known observations of the Haeffner effect. The quantitative disagreement is attributed to the choice of the form of the perturbing potential, the use of an electron mean free time, the uncertain character of the relation between self-diffusion and viscosity and the anharmonic character of the vibrations.

2.4 Theoretical Models

2.4.1 The theory of Fiks

Fiks[35,104], has examined electromigration in metals within the framework of the free electron model and the basic concepts of the kinetic theory of condensed phases. He has shown that the effective mobility is determined by the coefficient of diffusion and by the electrical properties of the metal, namely, the mean free path and scattering cross-section of the ions for electrons. It is considered that the "friction force", exerted by the electrons on the ions, is equal to the rate of electron to ion momentum transfer; on collision with an ion, electrons are assumed to lose the total additional momentum acquired from the accelerating field since the previous collision. This force is calculated by a rigorous quantum mechanical approach assuming a spherical Fermi distribution and completely degenerate electron gas, and is found to be

$$F_{ei} = eEn_e \ell A_i \quad , \tag{78}$$

where n_e is the number of free electrons per unit volume, ℓ the mean free path and A_i the electron scattering cross-section of the ion.

Since the field force on an ion of valency Z_i is $Z_i eE$, the resultant force, taking account of the electron-ion collisions is thus

$$F = eE(Z_i - n_e \ell A_i). \tag{79}$$

For a pure metal and ions in the ground state $n_e \ell A_i = Z_i$ so that $F = 0$ and the metal is in mechanical equilibrium, but for an impurity ion or one in an activated state, $n_e \ell A_i \neq Z_i$ and a net non-zero force results.

Modifying a relationship earlier established by Mott[36], Fiks attempted to estimate the electron-scattering cross-sections of impurities from their effect on the residual electrical resistivity of a solvent metal. An equation for the friction force :

$$F_{ei} = - eE. \frac{\Delta \rho_i}{\rho} . \frac{1}{N_i} \quad , \tag{80}$$

where ρ is the specific resistivity of the solvent metal and $\Delta \rho_i$ the residual resistance caused by the presence of atom fraction N_i of impurity, is obtained. However, since values of $\Delta \rho_i$ arise from the entire distorted regions of the lattice and not from the impurity ions alone, equation (80) is found to yield values of the friction force which are too large. Like Klemm[31,32] Fiks therefore postulates that migration involves the movement of activated ions of scattering cross-section A_i^* and true charge Z_i^*, different from those of normal ions. The total force on an i ion in the activated state is then given by

$$F_i = eE(Z_i^* - \bar{Z}\frac{A_i^*}{\bar{A}}) \quad , \tag{81}$$

where $\bar{Z}$ and $\bar{A}$ represent mean values for the components of the solution. The terms in parenthesis thus represent the effective valency.

2.4.2 Belashchenko's analytical treatments

A relationship similar to that of equation (81) was obtained by Belashchenko[37] directly from considering the condition for mechanical equilibrium of a specimen. Using the terminology of (78) and (79), and summing over all components, this condition may be represented by

$$\sum_i N_i(eEZ_i - F_{ei}) = 0 \ . \tag{82}$$

For a two component system, assuming F_{ei} is a linear function of the electric field so that $F_{e1}/F_{e2} = A_1/A_2$, it is then easily shown that the effective valences of the two components may be expressed as

$$Z_1^o = \frac{N_2(A_2Z_1 - A_1Z_2)}{N_1A_1 + N_2A_2} \tag{83}$$

and

$$Z_2^o = \frac{N_1(A_1Z_2 - A_2Z_1)}{N_1A_1 + N_2A_2} \quad , \tag{84}$$

while for dilute solutions, e.g. $N_2 \rightarrow 0$, equation (83) reduces to

$$Z_1^o = (Z_1 - Z_2 \cdot \frac{A_1}{A_2}) \quad . \tag{85}$$

By assuming that the scattering cross section is proportional to the square of the true valence, the above equation may be rewritten as

$$Z_1^o = (Z_1 - \frac{Z_1^2}{Z_2}) \quad . \tag{86}$$

Belashchenko[38-40] has found reasonable agreement between effective valences obtained experimentally, from steady state electromigration measurements on a number of dilute solutions, and values calculated from equation (86) using normal chemical valencies for the components. In a further empirical analysis, Belashchenko[37] points out that equation (83) may be converted to the form

$$Z_1^o = \frac{N_2(\frac{A_2}{A_1} Z_1 - Z_2)}{1 - (1 - A_2/A_1)N_2} \qquad (87)$$

or, assuming Z_1, Z_2 and A_2/A_1 are independent of composition

$$Z_1^o = \frac{aN_2}{1 - bN_2} \quad , \qquad (88)$$

where a and b are constants. Experimental values of Z_1^o, determined as a function of composition, may thus be used in conjunction with the last relation to estimate the chemical valencies and ratio of scattering cross-sections of the components in the alloy. By combining such electrotransport data with those for the resistivities of the liquid alloys he has also evaluated the mean free paths of the conduction electrons and the general character of the density of states at the Fermi level[105].

2.4.3 The theory of Bresler and Pikus

An expression for the friction force similar to that obtained by Fiks was derived in the very simple treatment by Bresler and Pikus[41]. Again assuming it to be determined by the rate of change of electron momentum on collision, they write for the electron-ion interaction force:

$$F_{ei} = \frac{m_e j}{n_i e \tau} \quad , \qquad (89)$$

where n_i here represents the number of ions per unit volume, j the electron flux, $m_e j/e$ the electron momentum per unit volume and τ the mean free time between collisions. Alternatively, since $E\sigma = j$, equation (89) may be written

$$F_{ei} = - \frac{m_e \sigma E}{n_i e \tau} \quad . \tag{90}$$

Modifying the classical formula for conductivity to allow for effective electron mass (m_e^*), $\sigma = n_e e \tau / m_e^*$, and putting $n_e/n_i = \bar{Z}$, the average valency, the above equation then yields

$$F_{ei} = - \bar{Z}eE\left(\frac{m_e}{m_e^*}\right) \quad . \tag{91}$$

It is then postulated that the total force acting on an ion in a pure metal is zero, so that the field force

$$F_{fi} = - F_{ei}$$

$$= + \bar{Z}eE\left(\frac{m_e}{m_e^*}\right) \quad . \tag{92}$$

The friction force on a given component is assumed to be directly proportional to the contribution of that component (ρ_i) to the total alloy resistivity (ρ); thus they write

$$F_{ei} = - \bar{Z}eE\left(\frac{m_e}{m_e^*}\right) \cdot \frac{\rho_i}{\rho} \quad . \tag{93}$$

Combining equations (92) and (93) thus suggests that the net force on the ion species i is

$$F_i = \bar{Z}eE\left(\frac{m_e}{m_e^*}\right)\left(1 - \frac{\rho_i}{\rho}\right) \quad . \tag{94}$$

Clearly severe approximations are involved in this treatment and, in particular, the assumption of zero net force on the average atom is of doubtful validity in the light of observations on self-transport.

2.4.4 The theory of Huntington and Grone

A viewpoint akin to that of Fiks has been used by Huntington and Grone[42] to calculate the electron friction force. They have examined the interaction between the charge carriers and <u>moving</u> atoms ("defects") and consider the momentum transferred to the atoms when at the saddle point between lattice sites as influencing the direction of jumping. The rate of such momentum transfer has been calculated using a semi-classical treatment, with the simplifying assumptions that the current carriers are scattered by the defects alone and that the defects may be considered to be decoupled from the lattice. In the original treatment the rate of momentum transfer was calculated quantum mechanically from the rate of change of electron momentum, where the latter was taken as m_e times the group velocity (V_g) of the Bloch wave. It was further assumed that the mean free time between electron-defect collisions (τ_d) was independent of wave number and argued that the total number of collisions occurring in jump time is sufficiently high to justify approximating the effect as a continuous force:

$$F_{ed} = -(j\, m_e)/(e\tau_d n_d) \quad ; \tag{95}$$

n_d is the density of defects and j the current density in the x-direction.

Expressing the contribution of the defects to the resistivity by

$$\rho_d = |m_e^*|/n_e e^2 \tau_d \tag{96}$$

$$\text{or} = |m_e^*|/n_i Z e^2 \tau_d \quad , \tag{97}$$

where the electron density (n_e) is replaced by the ion density times valency ($n_i Z$), thus yields from equation (95)

$$\begin{aligned} F_{ed} &= -n_i e j \rho_d m_e/(n_d |m_e^*|) \\ &= -eEZ \frac{n_i \rho_d}{\rho \cdot n_d} \cdot \frac{m_e}{|m_e^*|} \quad , \end{aligned} \tag{98}$$

This relation is analogous to equation (93) obtained in a less rigorous manner by <u>Bresler and Pikus</u>[41]. It would suggest, however, that the direction of the friction force is independent of the sign of the current carrier. This result is a consequence of using the average change of m_eV_g to calculate the change of momentum and in a subsequent paper Huntington[43] argues that this is erroneous and that the calculation of the rate of change of momentum should be based on averaging changes of the pseudo-momentum k. Thus since the ratio of k to m_eV_g is m_e^*/m_e, the corrected result for F_{ed} is

$$F_{ed} = - eEZ \frac{n_i \rho_d}{\rho . n_d} \cdot \frac{m_e^*}{|m_e^*|} \quad , \qquad (99)$$

whence a dependence of direction of the friction force on the sign of the charge carrier is evident.

The friction force is considered to be a function of microscopic position, being a maximum (F_{ed}) at the saddle point and a minimum at a lattice point. For simplicity it is assumed to vary sinusoidally with the periodicity of the lattice. The kinetics of the process are then treated by developing a modification of the Nernst-Einstein equation (7) appropriate for a particle whose charge varies with position. Averaging the chosen sinusoidal function over the migration paths results in modifying the normal Nernst-Einstein relation merely by a factor of 2 in the denominator.

The field force on an ion is assumed to be due to the direct electrostatic interaction of the field with the unscreened ion. Combining this with the above treatment of the friction force thus gives, for the total force acting on a moving ion:

$$F_i = eEZ \left[1 - 1/2 \; \frac{\rho_d n_i m_e^*}{n_d \rho \, |m_e^*|} \right] \quad . \qquad (100)$$

Here ρ_d is the contribution of the moving ions, considered as defects, to the resistivity of the sample and n_d/n_i is the relative concentration of such defects.

2.4.5 The theory of Bosvieux and Friedel

The most sophisticated quantum-mechanical treatment of electromigration in metallic alloys is that by Bosvieux and Friedel[44]. They have calculated the perturbation of the electron wave functions, produced by the application of a small field, and the effect of this on interstitials, vacancies and substitutional atoms. In addition to direct polarization effects, they have considered effects arising from the displacement of the Fermi distribution due to the passage of the electric current. A simplified treatment, based on a free electron model and neglecting electron correlations, is employed.

The direct action of the electric field is treated by considering its modification of the wave functions within the Fermi sphere. It is found that in the free electron approximation all electrons and ions feel the direct field without screening effect. It is further shown that a substitutional impurity feels the same force as the matrix atoms, while an interstitial impurity does not feel any direct force. Calculation of the polarization due to the displacement of the Fermi surface reveals the existence of a net force parallel to the applied field and proportional to its intensity. The force is a function of the Fermi level of the matrix and proportional to the collision relaxation time of the current carriers and hence to the electrical conductivity of the matrix; it thus varies inversely with temperature.

Qualitatively, the displacement of the Fermi surface may be considered as equivalent to a reduction of electron velocity on the anode side of the impurity and an acceleration on the cathode side; the electron density is thereby increased in the region where they are moving more slowly. Being more numerous, they give rise to stronger screening and an augmentation of charge which draws the impurity towards the anode. Bosvieux and Friedel's treatment shows that, while for an electronic conductor this force is anode directed, an exactly opposite expression resulting in entrainment of defects towards the cathode will arise in the case of a positive hole conductor.

For an interstitial impurity they show that, since the direct polarization effects exactly compensate the electrostatic force, the net force is simply

$$F_i = - eEZ \frac{n_i \rho_d}{n_d \rho} \quad , \tag{101}$$

where again Z is the valency of the matrix, ρ its total resistivity, n_d/n_i the concentration of defects and ρ_d the residual resistivity of these. This result is exactly equivalent to that obtained in Huntington and Grone's semi-classical treatment. Equation (101) thus shows that the net force on an interstitial ion is in the same direction as the carrier current and proportional to the valency of the matrix. Since $\rho_d n_i/n_d$ is independent of temperature, the force will be seen to decrease with increasing temperature as a consequence of its dependence on the conductivity of the matrix.

Self-transport arises from the movement of atoms into adjacent lattice vacancies in a manner producing net transfer. Bosvieux and Friedel's treatment of the case of vacancies thus provides a model for this phenomenon. Like Huntington and Grone[42], these workers also consider how the net force on a migrating ion will vary with position. The force on an ion when on a lattice site is once more calculated from the displacement of the Fermi surface, while it is assumed that the force when at the saddle point is identical with that shown above for an interstitial. These two values are then combined again by assuming a sinusoidal variation having the periodicity of the lattice. This yields, for the effective net force on an ion undergoing self-transport,

$$F_i = - 1/2 \ eEZ \ \{\frac{n_i \rho_d^{(z)}}{n_d^{(z)} \rho} - \frac{4 \Upsilon Z}{3\pi} \quad [J_o(b) + 2J_2(b)] -1 \} , \tag{102}$$

where $n_i \rho_d^{(z)}/n_d^{(z)}\rho$ is the value of the quantity $n_i \rho_d/n_d \rho$ for interstitials of charge Z, J_o and J_2 are Bessel functions and b the interatomic distance. Since the value of the net force is determined by the difference in magnitude between the terms within the outer brackets, a change of sign of the resultant force is possible as a result of the variation of ρ with temperature; it is suggested that this may account for the reversal of the direction of electrotransport which is sometimes observed.

The electromigration of substitutional solutes is treated in an exactly analogous fashion to that used for self-transport, since the diffusion mechanism will again involve movement into adjacent sites. For substitutional impurities of charge (Z + z) in a matrix of ionic charge Z, the expression for the effective net force becomes, using the notation of equation (102),

$$F_i = -1/2\ eEZ\ \{\frac{n_i}{\rho}\left(\frac{\rho_d^{(z)}}{N_d^{(z)}} + \frac{\rho_d^{(z+Z)}}{n_d^{(z+Z)}}\right) - \frac{4\pi Z}{3\pi}\ [J_o(b) + 2J_2(b)] - 1\ \}\ . \qquad (103)$$

As in the case of a pure metal, where the current carriers are electrons, the effective force is anode directed at normal temperatures but may change sign at very high temperatures. Clearly both the solute and solvent ions may be displaced towards the same electrode by this mechanism.

Measurements on alloys normally yield directly only the relative movements of the components. This movement is related to the differential force on the components, which may be written,

$$\Delta F = -eEZ\frac{n_i}{2\rho}\left(\frac{\rho_d^{(z)}}{n_d^{(z)}} + \frac{\rho_d^{(Z+z)}}{n_d^{(Z+z)}} - \frac{\rho_d^{(Z)}}{n_d^{(Z)}}\right)\ . \qquad (104)$$

In the approximation used, the quantity $n_i\rho_d/n_d$ is proportional to the square of the excess solute charge. Thus one may write

$$n_i\rho_d/n_d = \lambda z^2 \qquad (105)$$

and the differential effective force may then be expressed as

$$\Delta F = -\ eEZ\frac{\lambda}{\rho}Z(z + Z)\ \ . \qquad (106)$$

For an electronic conductor the effective relative force on the solute is thus in the same sense as the field when $Z<z<0$ and reversed for other values of z. The model is only valid for low concentrations of solutes or defects since interactions between impurities have been neglected. The results may be significantly modified by such interactions which will be present in concentrated alloys. Experimental verification of equation (106) has been provided in investigations, by Van Doan[78], of the electro-migration of Ag, Cd, In, Sn and Sb in silver. The results are in good accord with the theoretical model in showing the predicted linear dependence of the effective valencies on $Z(z + Z)$ and a decrease in the effective valencies with increasing temperature.

2.4.6 The theory of Gerl

Using the Bosvieux-Friedel model, Gerl[77,106], has analysed the force due to the modification of the Fermi distribution of the carriers in the matrix by making explicit calculations of the spatial distribution of electrons and the effect of perturbation by an external electric field. His results are again interpreted in terms of specific resistivities. Following the previous treatment he again assumes that an interstitial of excess charge z is completely screened so that the direct electrostatic force acting on the impurity ion vanishes and the only force therefore arises from the electron wind and is

$$F = - ZeE \left(\frac{\rho_i}{\rho_o}\right) \quad , \qquad (107)$$

where as before ρ_o is the resistivity of the pure matrix and ρ_i the specific resistivity of the solute ion; the latter is not exactly the same as that ($\rho_{i(Linde)}$) obtained from a resistivity measurement since $\rho_{i(Linde)}$ includes the scattering of electrons on the interstitial ion and its relaxed neighbours, whereas ρ_i is only due to the scattering on the ion itself. It is to be expected that $\rho_i \leqslant \rho_{i(Linde)}$ and the difference may be important for big interstitial ions whose neighbourhood relaxation is rather large.

In the case of a vacancy diffusion mechanism the situation is complicated by the presence of a neighbouring vacancy and the change of screening charge over the jump. For an impurity ion of charge Z + z in a matrix Z, the screening charge in the stable position is z and the force at this point is expressed by

$$F^{Z} = ZeE\ [(1 + f_o) - (\frac{\rho_i^{z}}{\rho_o})] \quad , \tag{108}$$

where f_o represents the force acting on the impurity ion due to the polarization of the screening charge of the neighbouring vacancy and is calculated in an approximation where the screening charges of the two defects are assumed to be independent of each other. During the jump, assuming that the direct electrostatic force vanishes and that the f_o term arising from the polarization of the two half-vacancies can be neglected, he obtains for the force at the saddle point

$$F^{z+Z} = -\ ZeE(\frac{\rho_i^{z+Z}}{\rho_o}) \quad . \tag{109}$$

Gerl then writes, for the average force along the jump,

$$\begin{aligned} F &= 1/2\ [F^{Z} + F^{z+Z}] \\ &= 1/2\ ZeE\ [1 + f_o - \frac{\bar{\rho}_i}{\rho_o}] \quad , \end{aligned} \tag{110}$$

where $\bar{\rho}_i = \rho_i^{z} + \rho_i^{z+Z}$.

Arguing as before that $\rho_i^{z} \leqslant \rho^{z}_{i(Linde)}$ and $\rho_i^{z+Z} \leqslant \rho^{z+Z}_{i(Linde)}$, it follows that

$$\bar{\rho}_i \leqslant (\rho_i^{z} + \rho_i^{z+Z})_{(Linde)} \quad . \tag{111}$$

Values of $\bar{\rho}_i$, the sum of the excess resistivities due to the ion at the stable position and at the saddle point, have been derived by Van Doan[78] from his results for the electromigration of Cd, In, Sn and Sb in silver. To check the above idea, Gerl has compared these with estimated $(\rho_i^{z} + \rho_i^{z+Z})_{Linde}$ values;

the latter have been calculated on the basis of the assumption that the specific resistivity of a z-valent ion at a saddle point is of the same order of magnitude as that of a (z + 1) valent substitutional ion, since both have the same screening charge. The comparison between the experimental values of $\bar{\rho}_i$ and the estimated values is fairly good, with the discrepancies being traceable to the approximations involved. Gerl suggests that this type of interpretation would be suitable for transition impurities in normal metals and for impurities in transition metals.

Van Doan[78,107] has discussed his values for $\bar{\rho}_i$ in a slightly different manner. By assuming that the excess resistivity of an atom in a stable position (ρ_i^z) will differ very little from experimental Linde values, the difference ($\bar{\rho}_i - \rho_i^z$) is taken to indicate the excess resistivities of the atoms when at the saddle points. The results of this analysis suggest that the difference between the stable and saddle point resistivities decreases as the valency difference between solute and solvent increases, so that for the case of Sb in silver the two resistivities are virtually identical. This implies that the electronic structure of Sb in silver remains almost the same in the two positions, while lower valent solutes undergo electronic variations as the ions pass from stable to saddle points. The electromigration of the transition metal solutes, Mn, Fe, Co and Ni in silver have also been investigated by Van Doan[84]. Comparison of experimental $\bar{\rho}_i$ values with Linde-value based estimates of ρ_i^z in this case suggests improbably low excess resistivities for all these solute ions at the saddle point; this anomalous result is attributed to the effect of vacancy flow on the apparent effective valencies in these alloys.

The three theoretical treatments of the wind force by Fiks, Huntington and Grove, and Bosvieux and Friedel, although different in approach, have yielded essentially the same result in that they relate the effective forces on the ions to the resistivity of the solvent and the excess resistivity of the solute or defect in the stable and saddle positions. This similarity is clearly displayed by rewriting their results in the simplified uniform fashion tabulated below:

	Average force on solvent ion of valence Z.	Average force on solute ion of valency Z+z
Fiks	$ZeE[1 - \frac{\rho^{Z}}{\rho_o}]$	$ZeE[1 - \frac{\rho^{(Z+z)}}{\rho_o}]$
Huntington + Grone	$1/2\ ZeE[2 - \frac{\rho^{Z}}{\rho_o}]$	$1/2\ ZeE[2 - \frac{\rho^{(Z+z)}}{\rho_o}]$
Bosvieux + Friedel	$1/2\ ZeE[(1 + f_o) - \frac{\rho^{Z}}{\rho_o}]$	$1/2\ ZeE[(1 - f_o) - \frac{(\rho^{z} + \rho^{(Z+z)})}{\rho_o}]$

where ρ^{Z}, $\rho^{(Z+z)}$ and ρ^{z} respectively represent the specific excess resistivities of the solvent ions at the saddle point, the solute ions at the saddle point and the solute ions at the stable point; ρ_o is the specific resistivity of the solvent. In each case, the second term within the main brackets is responsible for the electron wind effect while the first refers to the electrostatic force; f_o, it will be recalled, is a function characteristic of the solvent and takes account of the presence of a vacancy neighbouring the migrating ion.

An attempt to assess the relative merits of these three hypotheses, on the basis of data for the electromigration of Ag and Sb in silver, has been made by Van Doan and Brebec[85]. Their experimental results permit the determination of ρ^{Z}, ρ^{Z+z} or, in the case of the Bosvieux-Friedel relation, $(\rho^{z} + \rho^{(Z+z)})$ from the above models; these values are then compared with theoretical estimates for the same quantities, evaluated by the method of pseudo-potentials. The comparison is, unfortunately, inconclusive, although Van Doan and Brebec suggest that best accord is indicated with the Bosvieux-Friedel model.

2.5 Current Basic Problems

From the foregoing review it will be evident that although the principal factors involved in electromigration are generally recognised, rigorous and quantitative fundamental understanding is far from complete. There is fairly universal agreement that, in determining the driving force, momentum exchange is more important than direct interaction of the ions with the electrostatic field, but precise treatment of these forces remains difficult and uncertain. Some of the remaining basic problems and attempts to resolve them have been discussed recently by Huntington[45,86]. Amongst these is the still imperfectly resolved role of the direct electrostatic force and the extent to which, in this context, the ionic charge may be equated to the normal valency or whether it is so reduced by screening as to become effectively zero in many cases. Critical experiments involving electromigration in the presence of transverse magnetic fields[46,47] and careful comparison of electromigration and diffusion[48], may provide some evidence on the significance of the electrostatic force. The problems of properly defining this force, and of calculating it directly from first principles, have been surveyed in a recent paper by Frohberg[108]. Interstitial electromigration presents few anomalies, but further study is of particular interest since it is simpler than the substitutional case. Simplification arises because the drag force on an interstitial probably does not vary appreciably between equilibrium and saddle positions. It is suggested [45] that systematic studies of electromigration, diffusivities and contributions to residual resistivities of different interstitials in the same matrix may again serve to resolve the relative roles of electron wind and direct field forces.

Huntington[45] has also emphasised the importance of considering correlation effects in the case of substitutional impurities moving by a vacancy mechanism.

He has pointed out that these effects are considerably more complex than has been generally recognised, since neither the motion of the impurity nor that of the vacancies are uncorrelated. He has treated the particular case of the electromigration of an impurity in a f. c. c. lattice in terms of the standard jump frequencies (ω_1, ω_2, ω_3, ω_4) of a vacancy in the neighbourhood of an impurity; ω_1 represents the jump frequency of a vacancy about an impurity, ω_2 the frequency of vacancy-impurity interchange jumps and ω_3 and ω_4 the respective frequencies of vacancy-impurity pair dissociation and recombination jumps. It is assumed that the jump frequencies are infinitesimally altered by the applied field depending on the orientation of the jump direction to the field. This effect is simplified to involve only two parameters ε and σ, representing the strengths of electron drag on the vacancy and impurity respectively, and the final result for the jump frequency of an impurity in the direction of the applied is

$$\omega \omega_2 \frac{\sigma\{ 2\omega_1/\omega_3 + 5\} - 2\varepsilon \{2 \omega_1/\omega_3 - 3\}}{2 \omega_1 + 2 \omega_2 + 5 \omega_3} \qquad (112)$$

From the above result it may be concluded that

(i) where impurity-vacancy repulsion occurs (i. e. $\omega_3 > \omega_4$) the impurity flow will tend to be less than the flow of matrix atoms.

(ii) where the impurity attracts the vacancy, but does not exchange readily with it (i. e. $\omega_2 < \omega_1 , \omega_3 , \omega_4$), the impurity may tend to be drawn back by the vacancy and move counter to the mass flow,

(iii) where the impurity attracts the vacancy and exchanges readily with it, the flow is still not proportional to ω_2 alone but involves also a factor of the form $\omega_2(2 \omega_1 + 2 \omega_2 + 5 \omega_3)^{-1}$ so that the above expression approaches a limiting value of $1/2\, \omega \sigma (2 \omega_1/\omega_3 + 5) - 2\varepsilon(2\omega_1/\omega_3 - 3)$ as ω_2 goes infinite.

Huntington suggests that this final result is of significance in relation to the observation of very large effective valencies in some studies.

The role of vacancy flow in influencing solute electromigration has been examined by Van Doan[84,107], who has suggested that the presence of vacancy flow effects may result in "apparent" effective valencies which differ from the "true" effective valencies. As shown earlier, (Section 2.2.3), the electric mobility of a component in dilute solution, moving under the influence of a field, and by means of a vacancy flow mechanism, is given by

$$V_B = \frac{D_B^A Z_B^o}{kT} \left[1 + \frac{L_{BA}}{L_{BB}} \cdot \frac{Z_A^o}{Z_B^o} \right] e \quad , \tag{113}$$

where D_B^A is diffusion coefficient of B in solvent A, Z_B^o is the true effective valency of B, L_{AB} and L_{BB} are the usual phenomenological coefficients relating respectively to the cross-coupling between the fluxes of A and B atoms and to the diffusion of B; the remaining terms have their normal significance. Comparison of equation (113) with the form of the Nernst-Einstein equation showed that measured values of V_B will yield apparent effective valencies, Z_B^{oo}, where

$$Z_B^{oo} = Z_B^o \left[1 + \frac{L_{BA}}{L_{BB}} \cdot \frac{Z_A^o}{Z_B^o} \right] \tag{114}$$

In the second term in the bracket, a correction neglected in previous papers, the ratio L_{BA}/L_{BB} represents the "vacancy flow effect" and its value is determined by the values of the standard vacancy jump frequencies. Since the value of L_{BA}/L_{BB} may be positive or negative, enhancement or reduction of the true effective valency is possible. Van Doan's investigations suggest that this vacancy flow factor is responsible for the observation of low apparent effective valencies for the electromigration of the transition metal solutes, Mn, Fe, Co and Ni in silver. On the other hand, depending on jump frequencies, L_{BA}/L_{BB} may be small, so that the vacancy flow term becomes insignificant and the apparent and true effective valencies are identical; this is considered to be the case for the solutes Sb, Sn, Cd and In in silver. Unfortunately there are at present few systems where the necessary jump frequency data are available to enable rigourous application of the above analysis. The determination of vacancy jump frequency ratios in the neighbourhood of an impurity atom requires a knowledge of the enhancing effect of the solute on the solvent diffusion coefficient, of the ratio of the impurity diffusion coefficient to that of the solvent, and of the impurity correlation factor; where experimental measurements of the last factor from isotope effects are not available, Van Doan[110] has suggested the alternative use of measurements of marker velocities in electromigration in dilute solutions in the calculation of jump frequency ratios. Van Doan and Brebec[111] have recently proposed a general method for the derivation of the effective valencies of the components in non-dilute binary alloys. From equation (114) and the corresponding expression for the second component, it is readily seen that the true effective valency of A can be expressed as

$$Z_A^o = \frac{L_{BB}(L_{AA}Z_A^{oo}) - L_{AB}(L_{BB}Z_B^o)}{L_{AA}L_{BB} - (L_{AB})^2} \qquad (115)$$

and that an analogous expression can likewise be written for Z_B^o. The phenomenological coefficients L_{AA}, L_{BB} and L_{AB} may be evaluated in an approximate manner using equations derived by Manning[112] on the basis of a simple model which takes into account the vacancy flow effect. Using these together with values of Z_A^{oo} and Z_B^{oo}, derived from measurements of the total flux of the components and the marker shift in a homogeneous region of the specimen, then permits the calculation of the true effective valencies from the relations of the form of equation(115). These authors have used this method to re-evaluate the data for Ag-Au solid solutions reported by Hofman and Guy[113].

Attention is drawn to the importance of improved treatments of the basic problem of the calculation of the friction force from first principles. The most sophisticated theory existing at present, that of Bosvieux and Friedel[44], still employs an essentially free electron model and first-order Born approximation scattering. Clearly theoretical developments must be made to take account of the exact form of Fermi surfaces and the band structure in real metals; only then is reliable prediction of the friction force in complex cases involving holes and electrons likely to be achieved. More realistic treatment of scattering is also essential and it is believed that the possibility of progress in this may be offered by recent calculations using pseudo-potentials[49]. This approach has been employed in theoretical efforts to account for the dependence of electro-migration on orientation which has been observed in single crystals of zinc[50] and cadmium[51]. Although having essentially isotropic conductivity, the mass flow per unit field in these metals is substantially greater in the basal plane than along the c-axis. Attempts to understand this anisotropy of electrotransport in terms of a nearly free electron model and using established pseudo-potentials have, however, not been successful[49]; the analysis shows that, because of the size and position of the Brillouin Zone gaps, any treatment involving the assumption of a constant relaxation time will predict an anisotropy opposite to that observed. It is hoped that a study of the real, phonon-limited relaxation times of the electrons may resolve this contradiction.

Finally, an area which is receiving increasing attention, owing to its importance in relation to the failure of conductor stripes in integrated circuits, is that of electrotransport in thin films; the effect of such factors as grain boundary orientation, surface coating and solute additions on void formation are being examined[115,116,117,119]. Adam[118] has reviewed the possible mechanism of electrotransport on surfaces and interfaces.

3. Experimental Studies of Electrotransport

3.1 Experimental Methods

Published results of electrotransport studies of solid and liquid metals, and alloys and appearing in the literature before Spring 1972, are presented in the following tables. Much of the early work is of doubtful quantitative significance, but it is probable that in most cases the net direction of migration is correctly established.

For the studies of solid materials, samples in the form of wire or cylindrical specimens have generally been employed[12,13,86]. In the simplest technique the direct current through the specimen arising from the applied field provided the sole heating of the specimen[42,55] and consequently a temperature gradient existed between the hot central regions and the ends clamped in the electrodes; temperature distributions were either measured[42] or calculated[89]. Preferred arrangements, however, have employed auxiliary heating[90] to provide a uniform temperature throughout the length of the specimen. Single specimens or composite assemblies of several small cylinders[90,103], sometimes separated by inert metal foils, have both been employed. In alloy studies, various techniques may be used to determine the distribution of the components induced by electrotransport; these include normal chemical, spectroscopic or vacuum-fusion analysis[4], radioactive-tracer analysis[91,92], the observation of the migration of phase boundaries[55], microhardness[4] or electrical resistivity measurements[89] or the detection of p:n junction movements[93] in semi-conductor materials. Except in the case of rapidly moving interstitial solutes, the changes of concentration distribution by themselves reveal only relative movements and even the direction of the migration is indeterminate. To establish this and to obtain values for the individual transport numbers of the components they must be combined with the simultaneous determination of overall mass transport. This information may be obtained by the use of inert marker or weighing techniques; in the majority of cases transverse scratches[42, 94], microhardness indentations[55, 95], or inert metal markers[96] have been employed, but the deposition of a layer of a suitably immobile radioactive isotope[91] at interfaces between adjacent cylindrical specimens is an alternative method. Observation of the relative movements of the markers then permits the computation of the absolute amounts of material transported. With the surface-inscribed marker techniques (vacancy flux methods), the shape of the specimen is important in determining the exact relationship between the observed velocity of marker motion (V_m) and the velocity of the mass transport within the specimen (V_i). Penney[94] has shown that these are related by an "isotropy factor", α, so that $V_m = -\alpha V_i$. The value of α depends on the geometry of the sample; for a short specimen $\alpha \rightarrow 1$ (uniaxial transport), while for a long specimen $\alpha = 1/3$ (isotropic transport). To evaluate the isotropy factor observations of the total marker displacement must therefore be coupled with measurements of changes in transverse dimensions. No such corrections are involved in the isothermal isotope technique[91]. Marker techniques are of course employed in the measurement of self-electromigration in pure elements.

The main experimental problem in electrotransport studies of liquid metals and alloys arises from the fact that these materials are characterised by relatively high thermal conductivities and low viscosities so that convectional stirring readily occurs. Since electrotransport is a slow process, any resultant separation of the components is rapidly obscured by slight stirring. This effect is a common source of error in much early work.

To reduce convectional stirring, electrotransport experiments on the liquid state are invariably conducted by the passage of current through samples contained in some form of capillary[9,14,69,97]. The results of such experiments are influenced considerably by the shape and orientation of the capillary. The straight horizontal capillaries of relatively large bore, used by early workers[9], are particularly unsatisfactory, since any separation of alloy components produces a density gradient along the sample and in turn causes longitudinal convection as the heavier layers of the melt flow under the lower layers. In addition, the separation of the components produces a resistivity gradient and hence uneven Joule heating and this may also lead to further density gradients and convection unless an adequate heat sink is provided. Results obtained with relatively large bore capillaries are therefore probably only of qualitative significance unless effective precautions to prevent density induced convection are taken in designing the apparatus. Such simple devices as zig-zagging[9] or spiraling[98] of the capillary are likely to be inadequate. Some workers[99] have attempted to suppress convection by using straight vertical capillaries while taking precautions to ensure that the sample density decreases upwards. To do this the current direction is chosen so that the less dense component migrates upwards and an external heater is used to maintain a slight positive temperature gradient up the capillary. It has been suggested[14], however, that, since the d.c. heating will cause the temperature at the axis of a capillary to be slightly higher than at the walls, convection will occur in all but very small diameter vertical capillaries owing to the tendency of metal to rise along the axis and descend along the walls. Other possible mechanisms of convection which are independent of specimen orientation have also been suggested as sources of error. Thus, for example, it has been calculated that, at high current densities, even very small tapering of the container tube will cause mixing of the liquid by magneto-hydrodynamic convection[76]; this effect is indicated to be proportional to the fourth power of the current density and to the tenth power of the capillary radius. It is also considered that scattering of electrons at the boundary between the walls of the capillary and the molten metal can produce so-called electro-convection[100] in which melt moves in the same direction as the electrons along the axis of the capillary and in the opposite direction at the walls; such effects are believed to be significant if current densities greater than about 100 amp. cm^{-2} are employed[14]. To ensure quantitatively accurate results it is therefore necessary to carry out measurements using small cross-section capillaries and low current densities. Pyrex glass or quartz tubes of 1.0 to 0.5 mm or less diameter have most frequently been employed[69, 97], but the alternative use of very thin (0.025 mm x 0.3 mm) ribbon-type capillaries has proved very effective in suppressing convection[65, 101]; electrotransport cells of this type have been constructed from metal, mica and quartz[101] or boron-nitride and quartz[65]. Generally the distribution of the components after electro-transport in the liquid state has been determined after solidification of the material in the capillary. As with the solids, various standard analytical methods have been employed; these include chemical, radioactive tracer and spectroscopic analysis, micro-hardness measurements, quantitative metallography and in the case of the Haeffner effect, mass spectroscopy. In a few cases, techniques permitting the direct observation of rates of migration while in the liquid state have been evolved; the measurement of electrical conductivities[69, 101] and differential thermo-electric potentials[102], using probes set along the length of the capillaries, and the direct observation of phase boundary movement[65] have all been used.

3.2 Experimental Results

3.2.1 Self-electrotransport

While self-electrotransport can be observed in pure solid metals by marker or radioactive-tracer techniques it can not, for obvious reasons, be observed in pure liquid metals except for the special case of isotope separation in the Haeffner effect.

Self-electrotransport data (summarised in table A) are not extensive and are sometimes contradictory. Measurements have usually been performed on wire specimens heated by the passage of the direct current. Only in more recent experiments has account been taken of the often appreciable thermotransport (Soret effect) caused by the thermal gradients existing along the specimens in such techniques.

In most simple metals (e.g. Na, Cu, Ag, Au) and other electronic (n-type) conductors (e.g. Al, In, Sn, Ni) self-electrotransport occurs towards the anode. This is consistent with the expected predominance of the electron momentum exchange force over the field force. For metals with positive Hall coefficients (p-type conductors) the behaviour is more confused. Thus, while hole-conductors such as Fe, Co and W move towards the cathode in a manner attributable to the reinforcement of the field force by the cathode-directed carrier (hole) momentum exchange force, others (Zn, Cd, Pb) unexpectedly show anode-directed migration. In metals with complicated electronic structures the net wind force will depend on the difference between the products of the densities, mobilities and interactions with ions of both types of carrier-electrons and holes. It has been shown that even in p-type conductors the overall electron momentum transfer must be anode-directed[52]; thus it would appear that in Zn, Cd, Pb and etc. the activated ions acquire a larger momentum transfer from scattering electrons than from holes, while the reverse situation holds for Fe, Co and W.

Anomalous behaviour has been reported for Cu, Pt and Zr. Cu is said to show a reversal of transport direction at temperatures above approximately 1000°C. This was accounted for by postulating that the normal predominance of the electron momentum exchange force operated at low temperature, but that this was outweighed by the field force at high temperatures. The validity of this observation is doubtful, however, since such a reversal at high temperatures is not confirmed by other workers. It has been suggested that this apparent behaviour may have been caused by void formation; the latter phenomenon has been observed in silver. The cathode-directed transport of Pt is exceptional for a n-type conductor, but is well substantiated. Since the electron wind must be anode directed, it would seem that the field force must predominate in the self-electrotransport of Pt. Very small effects have been observed in the case of self-transport in Zr and it would appear that a very close balance exists between the electrostatic force and the net wind force from holes and electrons. Possibly this is delicate enough to be reversed by dilute impurities or sub-structural variations, which may account for the conflict of the migration directions reported in the two investigations of this metal.

An examination of the collected self-electro-transport data reveals that, where a range of values is reported, the effective valence in general decreases with increasing temperature, while the velocity of transport increases in a manner consistent with a thermally activated process of Arrhenius type and activation energy similar to self-diffusion. In the cases of Ag, Au, Cd, Zn, Sn and Pb a linear relationship between migration velocity and current density has been demonstrated in accordance with theory.

Table A. Self-electrotransport in Pure Solid Metals

Metal	Direction of Migration
Li	Anode
Na	Anode
Cu	Anode (Disputed reversal at ca. 1000°C)
Ag	Anode
Au	Anode
Zn	Anode
Cd	Anode
Al	Anode
In	Anode
Ge	Cathode
Sn	Anode
Pb	Anode
W	Cathode
Mo	Cathode
Fe	Cathode
Co	Cathode
Ni	Anode
Pt	Cathode
Ti	Anode
Zr	Anode/Cathode (Very small effects)
U	Anode

3.2.2 The Haeffner Effect

Since Haeffner first separated the isotopes of liquid mercury by the passage of a direct current, comparatively few further studies of this phenomenon have appeared in the literature. Many of the early measurements were undoubtedly marred by the effect of the unexpectedly large current-induced convection. The early work has been critically discussed and re-appraised by Lodding[53], who has carried out many of the later more significant studies.

In all liquid metals investigated so far, the lighter isotope has accumulated at the anode. Klemm has recognised the effect as a particular form of the phenomenon of self-transport. The light isotope diffuses faster than the heavy one and will thus concentrate in the direction of net diffusive mass flow, whatever the cause of the latter. Epstein assumes that the isotopes in a pure liquid metal have identical electron scattering cross-sections, but that electrons impart a slightly greater part of their energy to the lighter isotope on impact. On the other hand, as described earlier, de Gennes suggests that the lighter isotope has a larger vibration amplitude and is therefore subjected to a larger electron drag force. Published work on the solid state is confined to a single experiment on uranium and again the light isotope accumulated at the anode. Lodding[53] has reported unpublished work on solid indium.

3.2.3 Electrotransport of Interstitial Solutes

During electrotransport, the velocity of an interstitial atom is about two orders of magnitude greater than that of the lattice ions. Therefore the direct mobility of an interstitial ion may be determined, rather than a differential or relative mobility. The quantitative results of experiments of the electrotransport of interstitial solutes have been summarised and discussed previously by Seith[11], Heumann[12], Verhoeven[13] and by Oriani and Gonzalez[54]. The data

Table B. Interstitial Electromigration

Solvent Metal	Anode-directed solutes	Cathode-directed solutes
α - Fe		H, D, C, N
γ - Fe	N	H, C, B
Fe - Cr		C
Fe - Ni		C
Fe - Mn		C
Fe - Si		C
Ni		H, D, C
Ni - Cr		C
Co		C
V		C, N, O
β - Ti	O	C
Zr	O	C
Pd		H
W		C
Ta		C
Y	C, O, N, H	
Th	C, N, O	
Pu		C, B
Lu	C, N, O	
Ce	O	
Gd	O	
Ag	O	

are included in section A of the compilation, but the main features are also presented in table B.

The total amount of work in this field is still comparatively small. Indications from existing data are that in a particular matrix all interstitial solutes go in the same direction and that this in general corresponds to the direction of movement of the effective charge carriers, be they electrons or holes. Thus the movement of H, D and C in α-Fe and γ-Fe and Fe alloys, C in Co, C, N and O in V and C in β-Ti are consistent with the assumption that cathode-directed defect electron (hole) momentum exchange is predominant; the high value of the effective valency observed for C in Fe suggests that this force may be augmented by a similarly directed field force acting on positively charged C ions. Similarly the movement towards the anode of O in Ag, C, O, N and H in Y and C, N and O in Th may be attributed to the influence of a normal anode-directed electron drag force.

The behaviour of N in γ-Fe appears anomalous when compared with that of all other interstitial solutes in this material, but the value of the effective valency obtained is quite small and, in view of the unsophisticated technique used in this instance, an error of sign is not inconceivable. The movement of O towards the anode in Zr and β-Ti also appears anomalous since these metals are p-conductors; if the carrier momentum exchange force is thus cathode-directed, the movement of the interstitial ions towards the anode would have to be due to the dominance of the electric field force, also implying that these particular interstitial ions must be present in a negatively charged state. Similarly Pd is predominantly n-conducting leading to the expectation of an anode-directed carrier momentum exchange force; the movement of H in the direction of the cathode is thus anomalous and would in this case imply that the interstitials are positively charged and that the direct field force is predominant. Such influence from the field force is , however, not consistent with the Bosvieux-Friedel conclusion that interstitials are not subject to a direct electrostatic force from the applied field. Perhaps the most puzzling behaviour, however, is that of the interstitials H, D and C in Ni and C in Ni-Cr alloys, since their motion towards the cathode conflicts completely with the n-type character of this solvent metal and with the well substantiated anode-directed self-migration of nickel. These particular observations thus raise doubts concerning the true direction of the force of interaction between the carriers and ions and suggest the possibility that it could have opposite signs for substitutional and interstitial migration. In considering the probable nature of the predominant momentum exchange in the transition metals, it must be remembered that the electronic structures are complex, frequently not perfectly characterised, and may be modified by the addition of the solutes themselves if these are present in significant concentrations.

Examination of the electrotransport data for interstitial alloys shows that in every case where a range of values is reported the electric mobility or transport number of the solute increases with increasing temperature. However, while the effective valency shows a decrease with increasing temperature in some systems, in others it either increases or is not regularly dependent upon temperature.

3.2.4 Electrotransport in Substitutional Solid Solutions and Intermediate Phases

Since in substitutional solid solution and similar structures movement of all components can be expected to take place, many measurements have merely indicated relative movements and so yielded only differential mobilities and transport numbers. Most more recent studies, however, have made use of marker indentation or radioactive tracer techniques so that true partial values for the individual components can be obtained. Pioneer studies of this type were those of Seith and Wever[55]; their results have been summarized by Heumann[12] who has made similar investigations. A large proportion of the published work has originated in the Soviet Union where Kuz'menko and his associates have investigated the mobilities of components in a number of copper, silver, zinc and aluminium alloys, while Frantsevich, Kalinovich and various co-workers have been principally responsible for studies of various intra-transition metal alloy systems; the latter workers have generally reported effective valencies, but in some instances, using a two band model for the metals, have attempted to derive the true valencies of the components from the temperature or composition dependence of these.

The first determination of absolute transport numbers by Seith and Wever[55] was particularly noteworthy since it demonstrated that both components moved towards the same electrode. Their further observation that the direction of migration was towards the anode in electronically conducting phases - α(Cu-Al), β(CuAl), γ(Cu-Sn), α(Ag-Zn) and β(Ag-Zn), but became cathode directed in defect conductors - γ(CuAl), δ(Cu-Sn), γ(Ag-Zn), demonstrated very clearly the importance of electron- or hole-wind forces. An unexpected effect, however, was the decrease of transport number with increasing current density in β(CuAl) and Ag-Zn alloys.

Partial rates for Ag and Zn and total transport rates in Ag-Zn alloys were also examined by Frantsevich et al. They confirmed that Ag and Zn atoms in all cases migrated towards the anode, but made the interesting observation that, while in the α-phase the rate of transport of silver is greater than that of Zn, the ratio of the rates is reversed in the β-phase. It was suggested that this must be due to a change in the relative electron scattering cross-sections of Ag and Zn between the two phases because it is clear that the electron wind force prevails in the transport of both components in both phases.

In the work of Kuz'menko et al the predominance of the electron wind over the field force is again clearly demonstrated. The scattering cross-sections and effective charges of migrating (i.e. activated) ions have been determined; it was concluded that effective charges on activated ions are less than for those oscillating normally on lattice sites, possible due to the variation of electronic structure or the screening of the activated ions by the valence electrons in the vicinity.

Electrotransport studies of the β-phase of the Ni-Sb system by Heumann and Stuer have indicated that only Ni is transported and that towards the anode. The rapid migration of Ni, even at low current densities is attributed to the presence of structural vacancies in the phase.

From the fairly extensive electrotransport data for solid solutions and intermediate alloy phases it can be seen that, in every case where a sufficient range of values is reported, the electric mobility or transport number of the solutes again invariably increases with temperature. With a few exceptions, the effective valency tends to decrease with increasing temperature.

The first study of electrotransport in a non-metallic material appears to be that of Kubaschewski and Reinartz on the phase Mg_3Bi_2. Only relative transport numbers were obtained but the observed net movement of Bi towards the anode and Mg to the cathode are as would be expected from the heteropolarity of the bonding if the field force is dominant. Subsequent studies have been largely concerned with the migration of solutes in dilute solution in semiconducting elements. In most cases evaluation of the data has lead to the conclusion that, at least at low temperatures, the electron-ion friction force is negligible and that the solute elements exist as positively charged ions which are transported towards the cathode by the field force; in some instances, however, the observed mobilities do reflect the influence of friction forces.

While, because of the low concentration of current carriers, the field force would be expected to be dominant at low temperatures, the increase in the number of mobile electrons on raising the temperature, or deviating from stoichiometry in compounds, may be expected to lead to friction forces becoming increasingly significant and possibly over-riding the field force. It is noteworthy that at high temperatures the observed net directions of migration are, with the sole exception of the case of Bi in Si, everywhere in accord with those which would ensue from an anode-directed electron wind and the probable relative scattering factors of the components (i.e. they accord with the Epstein-Paskin criterion discussed below). Particularly interesting in this context are the reversals of migration observed with In and Sb in Ge and with Au in Si on increasing the temperature. The movement of In towards the anode and Sb to the cathode, observed at low temperatures, are consistent with the action of the field force on an acceptor (In) and donor (Sb) solute respectively, while the relative scattering factors of In, Sb and Ge would indicate preferential movement of In to the cathode and Sb to the anode by an electron friction force, which accords with the high temperature behaviour. Similarly the movement of Au in Si to the cathode at lower temperatures and the anode at higher temperatures is consistent with the existence of a positively charged ion and a change from field to friction force dominance as the temperature is increased. Fiks has attributed these changes of direction to the temperature dependence of the electron scattering cross-section and has estimated inversion temperatures of approximately 1325°C for Si and 725°C for Ge. The decrease in relative transport of the components of Mg_3Bi_2, observed with increasing temperature and deviations from stoichiometry, may also reflect the increasing influence of the electron wind force since in this instance the relative migrations induced by this force may be expected to be opposite to those arising from the dominant field force.

A few measurements have been made on solute additions in semi-conductor compounds (e.g. Li, Cu, Zn, Cd, in InAs and GaAs). Behaviour is again consistent with the solutes existing as positive ions and migrating mainly under the influence of the field force, but with effective charges usually reduced below the normal values by the effect of the electron drag force. In heavily doped p-type GaAs, however, Cu is reported to migrate with a charge very close to unity, it being suggested that in this material the electron drag is almost balanced by the hole drag force acting in the opposite direction.

3.2.5 Electrotransport in Liquid Alloys

Although measurements have been made on a very large number of liquid alloy systems, few have been investigated comprehensively, for example in relation to the effects of temperature and composition. Naturally most of the investigations have concentrated on low melting point alloys. The earliest studies were those of amalgams by Skaupy and by Lewis et al and extensive studies of these were later made by Kremann[56]. More recently the behaviour of dilute solutions of many elements in liquid alkali metals and other low melting metals such as Cd, Sn, Tl, In, Pb and Bi has been the subject of systematic interest to such workers as Verhoeven and Epstein in the U.S. and Drakin and Belaschenko in the U.S.S.R. In comparison, far fewer studies have been made of more concentrated alloys, but significant contributions in this area have been made by Romadin and again by Kremann and by Belaschenko. Early work has been summarized by Schwarz[9] and a useful discussion of studies of electromigration in liquid alloys up to 1965 has been presented in the review by Belaschenko[14]

Since only relative migration rates can be determined in studies of liquid systems, the experimental data have usually been analysed merely in terms of effective valencies. Belaschenko, however, has also attempted empirically to derive the true valencies and ratios of scattering cross-sections of the components from his studies of electrotransport as a function of composition. While the results of most past studies have been treated phenomenologically, current programmes of investigation increasingly aim to provide systematic data relevant to the test and development of theoretical models; they frequently incorporate diffusion and electrical resistivity measurements.

Although a range of values is reported for very few liquid alloy systems, the limited data available indicate that in general there is a tendency for the effective valency of the solute to decrease and for the electric mobility to increase with increasing experimental temperature. Since they are liquid at room temperature, amalgams have until recently been the most frequently studied alloys. It is particularly interesting to note that independent workers have observed a reversal from anodic to cathodic migration of the solutes to occur at low concentrations of Na, K or Ba in Hg; there is, however, some disagreement as to the critical concentrations. These effects may reflect a variation of scattering cross-sections or possibly the existence of short range order in the liquid amalgam; in the solid state a series of intermetallic compounds occur in these systems. The room temperature resistivity isotherms of these liquid alloys have maxima at similar low concentrations, possibly due to the association of unlike atoms in the melt. It is conceivable that the reversal of migration may result from a variation of the drag force with the tendency to cluster formation. In some alloy systems it is possible that reversal of migration with composition may account for the apparent disagreement between different investigations.

Although many of the models outlined earlier permit the empirical analysis of experimental electrotransport data to yield values for effective cross-sections or valencies, their present use for the a priori prediction of the precise behaviour of particular systems is limited by the frequently complex or indeterminate nature of the information required. Consequently, repeated attempts have been made to establish criteria for electrotransport behaviour which are based on more readily determinable parameters.

Skaupy[16], who was the first to recognise the existence of the electron momentum exchange force, suggested (1914) that, since the field and drag forces were related to the resistivity of a metal, ions which increase the resistivity should accumulate at the anode, while those reducing the resistivity should gather at the cathode. Later, Kremann[56] (1926) suggested that in metallic solutions the component of lower ionisation energy should accumulate at the cathode, while in 1959, after reviewing the then existing data, Angus, Verhoeven and Hucke[57] concluded that the alloy component with the lowest atomic weight became concentrated at the cathode. These early hypotheses, reasonable when proposed, have in turn been invalidated as completely general criteria by subsequent experimental data.

More recently, however, a promising semi-quantitative criterion was proposed by Epstein and Paskin[58]. Since it has become increasingly apparent that the electron drag force is probably the dominant factor in most electrotransport behaviour, they have suggested that, in a liquid metallic alloy, the ionic species with the highest electron scattering cross-section should be influenced to the greatest extent by momentum transfer and so preferentially transported to the anode. The degree of separation of the alloy components would be proportional to the product of electron current density and the difference in scattering cross-sections of the component ions. Belaschenko[37-39] has obtained fairly good correlations by assuming scattering cross-sections to be proportional to the squares of the true valencies, but Epstein and Paskin have argued, on the basis of the Faber and Ziman[59] model, that the difference between the scattering cross-sections will be proportional to the difference in electrical resistivities of the pure liquid components. Thus the elements with greatest resistivity may be expected to accumulate at the anode.

Epstein and Paskin demonstrated their resistivity criterion by comparison with the experimental results for a series of solutes in liquid bismuth and mercury, but the present authors[60] have examined its validity more fully by reference to the wider variety of alloy data provided by the present literature survey. Systems considered are confined to those where the direction of migration does not undergo reversals with increasing temperature or solute concentration. These systems, with the differences in the resistivities of their components at their melting points and the reported directions of solute migration are shown in Table C; following the suggestions of Ziman[61] and Epstein[58,62], the resistivities of the monovalent metals have been multiplied by a factor of 2.5 to compensate for the difference in structure between the monovalent and polyvalent metals.

Examination of Table C reveals that, of the 123 examples listed, only eleven alloy systems are in definite conflict with the Epstein-Paskin hypothesis. Five of these, Bi-Co, Bi-Ni, Sn-Mn, Fe-Ni and Ni-Fe, involve transition elements and their behaviour is readily understandable since, as Belaschenko[14] has pointed out, transition metals may be expected to have anomalously large scattering cross-sections as a consequence of their unfilled d-shells. Liquid resistivity data are not available for the components in a number of systems, but for most of these cases the reported electrotransport behaviour is in accord with the anticipated relative magnitudes of the scattering factors.

Table C. Electrotransport in Liquid Alloys

System	Δρ	Solute Accumulates	System	Δρ	Solute Accumulates	System	Δρ	Solute Accumulates
Na - Sr	*	N	Hg - Li	- 31	C	In - Si	+ 47	A
- Ba	+ 111	A	- Cs	0	A**	- Ge	+ 38	A
- Ag	+ 19	N	- Mg	- 64	C	- Sn	+ 15	A
- Cd	+ 10	A	- Ca	*	C	- Pb	+ 62	A
- Hg	+ 67	A	- Ag	- 48	C	- Sb	+ 81	A
- In	+ 9	A	- Au	- 13	C	- Bi	+ 96	A
- Tl	+ 49	A	- Zn	- 54	C			
- Pb	+ 71	A	- Cd	- 57	C	Ge - Al	- 47	C
			- Ga	- 65	C	- Ga	- 45	C
K - Hg	+ 59	A	- In	- 58	C	- As	*	A**
- Tl	+ 41	A	- Tl	- 18	N			
- Pb	+ 63	A	- Sn	- 43	C	Sn - Mn	- 8	A**
			- Pb	+ 4	N	- Co	+ 54	A
Cu - H	*	C	- Bi	+ 38	A	- Ni	+ 37	A
- Sn	- 4	A**				- Cu	+ 4	C**
- Si	+ 28	C**	Al - H	*	C	- Au	+ 30	A
- Ge	+ 19	A	- Ag	+ 19	A	- Zn	- 11	C
			- Au	+ 54	A	- Al	- 24	C
Ag - Sn	+ 5	A	- Si	+ 56	A	- Ga	- 22	C
- Si	+ 37	A	- Ge	+ 47	A	- Tl	+ 25	A
- Ge	+ 28	A				- Ge	+ 23	A
			Ga - Hg	+ 65	A	- Sb	+ 66	A
Au - Si	+ 2	C**	- Si	+ 54	A	- Bi	+ 81	A
- Ge	- 7	C	- Ge	+ 45	A			
			- Sn	+ 22	A	Pb - Co	+ 7	A
Cd - Co	+ 68	A	- Bi	+103	A	- Zn	- 58	C
- Ni	+ 51	A				- Cd	- 61	C
- Ag	+ 9	A	In - Co	+ 69	A	- Tl	- 22	C
- Au	+ 44	A	- Ni	+ 52	A	- Sn	- 47	C
- Tl	+ 39	A	- Ag	+ 10	A	- Sb	+ 19	A
- Sn	+ 14	A	- Tl	+ 40	A	- Bi	+ 34	A
- Bi	+ 95	A				- Se	$+10^7$	A

Table C. Continued

System	Δρ	Solute accumulates	System	Δρ	Solute accumulates	System	Δρ	Solute accumulates
Sb - Ag	- 71	C	Bi - Zn	- 92	C	Co - Si	- 22	C
- Au	- 36	C	- In	- 96	C	- Ge	- 31	C
- Zn	- 77	C	- Tl	- 56	C			
			- Sb	- 15	C	Ni - Mn	- 45	C
Bi - Mg	- 102	C	- Se	$+10^7$	A	- Fe	+ 39	C**
- U	*	A	- Te	+ 420	A	- Si	- 5	C
- Zr	*	A				- Ge	- 14	C
- Cr	*	A	W - Mo	*	A			
- Fe	± 10	C				Pd - Si	*	C
- Co	- 27	A**	Fe - C	*	C	- Ge	*	C
- Ni	- 44	A**	- S	*	C			
- Pd	*	A	- Mn	- 84	C	Pt - Si	*	C
- Cu	- 77	C	- Ni	- 39	A**	- Ge	*	C
- Ag	- 86	C	- Si	- 44	C			
			- Ge	- 53	C			

Notes:

A - B = Solvent - Solute

Δρ = resistivity of solute - resistivity of solvent ($\mu\Omega$ cm.)

A = anode. C = Cathode. N = no effect.

* = resistivity data unavailable. ** = contrary to model.

Thus the accumulation of U, Zr, Cr and Pd at the anode, when in Bi, is consistent with the higher scattering factors of transition elements as mentioned above. The light elements H, C, S are invariably found to concentrate at the cathode - behaviour which can be attributed to the expected small scattering cross-section of these elements. This may indeed result in the cathode-directed field force being predominant on these solutes, in addition to preferential movement of the solvent to the anode by the momentum transfer mechanism. The failure of the Au-Si, Cu-Sn and Hg-Cs systems to conform to the hypothesis may be simply a reflection of the uncertainty attached to small $\Delta\rho$ values or may be due to the importance of other factors when small differences in scattering cross-section are involved.

The Epstein-Paskin hypothesis was proposed for electrotransport in liquid systems. It is interesting, however, using the present collected data, to examine its applicability to electrotransport in solid alloys. Since solid state resistivities may be complicated by zone effects and variations of lattice vibration, the melting point resistivities of the liquid components are again assumed to provide the most convenient parameter for the scattering cross-sections of the component ions. Electrotransport directions in solid alloys are compared with the corresponding $\Delta\rho$ values in Table D. Of the 67 examples, nine appear to behave in a manner contrary to the hypothesis; several of these are solid solutions based on the noble metals. Unfortunately, resistivity data are again lacking in a large number of cases, but in most of these the electrotransport behaviour is as would be anticipated from the foregoing discussion of the liquids.

By considering the behaviour of various solutes in liquid bismuth and mercury, Epstein and Paskin have also attempted to relate the magnitude of the differential electromigration in a liquid alloy to the solute-solvent

Table D. Electrotransport in Substitutional Solid Solutions

System	$\Delta\rho$	Solute accumulates	System	$\Delta\rho$	Solute accumulates	System	$\Delta\rho$	Solute accumulates
Cu - Fe	+ 72	A	Si - Fe	+ 44	N**	Co - W	*	C
- Co	+ 50	A	- B	*	C			
- Ni	+ 33	A				Ni - Cr	*	C
- Au	+ 26	A	Ge - Li	- 11	C	- W	*	C
- Al	- 28	A**	- Cu	- 19	C	- Mo	*	C
- Sn	- 4	A**						
- Sb	+ 62	A	Pb - Ag	- 52	A**	U - Fe	*	A
			- Au	- 17	A**	- Sn	*	A
Ag - Zn	- 6	A**						
- Sb	+ 71	A	Bi - Sb	- 15	C	Pu - Fe	*	A
						- B	*	C
Au - Pd	*	C	Se - Tl	-10^7	C			
- Cu	- 26	C				Ce - Mn	*	A
- Ag	- 35	A**	Te - Tl	-475	C	- Fe	*	A
- Sb	+ 36	A				- Co	*	A
			Mo - Cr	*	C	- Ni	*	A
Zn - Ag	+ 6	A	- Fe	*	C	- Mo	*	C
			- Ni	*	A	- Zr	*	N
Al - Ag	+ 19	A	- W	*	C	- Mg	*	N
- Zn	+ 13	A				- Si	*	N
			W - Fe	*	A	- Sb	*	N
Si - Li	- 20	C	- Mo	*	C			
- Cu	- 28	C	- Th	*	A	Y - Mn	*	A
- Ag	- 37	C				- Fe	*	A
- Zn	- 43	C	Fe - Cr	*	C	- Co	*	A
- Al	- 56	C	- Ni	- 39	N**	- Ni	*	A
- In	- 47	C	- W	*	C	- Ti	*	A
- Bi	+ 49	C**	- Mo	*	A	- B	*	A
			- Al	-100	C			

Note: Symbols as for Table C.

resistivity difference. A semi-quantitative correlation was obtained by using a coefficient of electron drag (P), rather than electric mobility, to express the electromigration data; P is defined as $n_Y/N_Y n_e$ where n_Y is the number of atoms of a component Y passing a fixed plane per unit time, N_Y the atom fraction of Y in the alloy and n_e the number of electrons which passed through the alloy per unit time. The use of this parameter (P) recognises that it is the electric current, rather than the electric field, which produces the relative motion. Correlation is good for polyvalent solutes in the polyvalent solvents, but deviations have been found, for example, in the behaviour of monovalent solutes (Cu, Ag) in liquid bismuth and of polyvalent solutes (Cd, In) in monovalent sodium. Unfortunately the present survey has yielded very few additional mobility data; while these are reasonably consistent with the suggestion that the differential mobility should be proportional to the scattering factor differences, the quantitative transport data remain inadequate, in number and accuracy, to establish a more exact correlation.

Examination of all existing electrotransport data supports the general validity of the Epstein-Paskin criteria to a remarkable degree, so that the elemental resistivity difference appears to provide a fairly reliable basis for the prediction of the qualitative electrotransport behaviour in many solid and liquid alloy systems. Nevertheless, the use of electrical resistivities of the elements at their melting points as a mesure of scattering cross-sections in alloys is obviously an oversimplification, since the screening of the ions and hence their effective scattering cross-section may vary with the alloy environment if the electron density in the alloy differs from that of the pure components. The simple model cannot, for example, account for reversals of the direction of migration, such as are observed in the liquid sodium-potassium alloys. More recently, therefore, Epstein and Dickey[80] have attempted to refine the simple empirical model by means of theoretical calculations of electron-ion coupling. They have estimated the scattering power of ions in Na-K liquid alloys by using the scattering phase shifts of an electron at the Fermi level calculated by Meyer et al[81].

The probability of deflection of an electron through an angle θ is calculated from the scattering amplitude $f(\theta)$,

$$f(\theta) = \frac{1}{2ik} \sum_{\ell} (2\ell + 1)(e^{2i\eta_\ell} - 1) P_\ell(\cos\theta) \quad , \qquad (116)$$

where ℓ is the orbital quantum number, η_ℓ the corresponding partial wave shift evaluated at $k = k_F$, k being the wave vector and k_F its Fermi level value. The largest transfer of electron momentum to an ion will occur when the direction of motion of the electron is reversed; Epstein and Dickey therefore consider the probability of back-scattering as a measure of the tendency of an ion to migrate with the electron stream. For back-scattering, $\theta = \pi$ and $P_\ell(\cos\theta) = (-1)^\ell$; the scattering cross-section, S, is then given by

$$S = | f(\theta) |^2$$

$$= \frac{1}{(2k)^2} \left[\left[\Sigma_\ell (-1)^\ell (2\ell + 1) \sin 2\eta_\ell \right]^2 + \left[\Sigma_\ell (-1)^\ell (2\ell + 1)(1 - \cos 2\eta_\ell) \right]^2 \right] \qquad (117)$$

Using $k = k_F = (3\pi^2 n)^{1/3}$, where n is the number of electrons per unit volume of the alloy and interpolating phase shift (η_ℓ)values from those calculated by Meyer et al, Epstein and Dickey have calculated the scattering cross-section of both Na and K ions as a function of alloy composition.

The results indicate that in nearly pure sodium, K will go to the anode, while in nearly pure potassium it will go to the cathode; a reversal in behaviour is thus predicted at intermediate compositions, agreeing qualitatively with that found experimentally. This treatment of liquid Na-K alloys has been extended by Olson, Blough and Rigney[82] who have computed the back-scattering cross-sections of Na and K as a function of composition, temperature and pressure. Recently the latter authors have made similar calculations pertaining to electrotransport in the 21 liquid binary systems involving the monovalent elements Li, Na, K, Rb, Cu, Ag and Au[83]. Their results indicate a wide variation of electrotransport coefficients and the reversal of electrotransport direction at critical compositions in a number of cases; such cross-overs are predicted in Li-K (48% K), Li-Rb (~ 100% Rb), Na-K (64% K), Na-Rb (75% Rb), Cu-Ag (17% Cu), Cu-Au (46% Au), K-Cu (44% K), K-Ag (45% K), K-Au (45% K), Rb-Cu (48% Rb), Rb-Ag (49% Rb) and Rb-Au (49% Rb), while no reversal is expected in the alloys of the remaining combinations. For the few systems for which experimental data are available, qualitative agreement with the model is good, although there are considerable discrepancies between the experimental and predicted values of the reversal compositions and the magnitudes of the relative drag coefficients. However, the phase shift approach appears to offer a promising method for the description of electrotransport behaviour and one which in principle may be extended to ternary and higher order alloy systems. The phenomenon of reversal of direction of electromigration has also been discussed by Belashchenko[105] and by Sinha[114]; in both cases it is shown that in binary metallic alloys, involving elements of the same valency, reversal is to be expected at the composition where the composition dependence of the resistivity ($d\rho/dN$) equals zero.

4. Applications of Electrotransport

In principle the separation of alloy components effected by electrotransport offers the possibility of its use as a purification technique in either the solid or liquid state. The problems of purifying metals by this means have been considered in some detail by Verhoeven[5] who has concluded that the degree of purification possible is basically determined by the value of the ratio -UEL/D, where L is the specimen length, E the electrical field intensity, U the differential electric mobility and D is the diffusion coefficient. Since purification is normally concerned with dilute solutions, U effectively reduces to the electrical mobility of the solute. It will be evident from this expression that, for a given experimental arrangement, the optimum temperature for purification will be determined by the temperature dependence of U and D. The variation with temperature of the differential mobility is considerably less than that of the diffusion coefficient so that U/D decreases with temperature and hence the amount of purification theoretically attainable will be higher at lower temperatures. Against this, however, must be balanced the fact that the time required to achieve this will increase as the diffusion coefficient becomes smaller. Verhoeven's calculations suggest that for values of U/D greater than about 12 significant purification can be achieved in a reasonable time (e.g. 5 days) under feasible experimental conditions.

Considered as possible, but not necessarily optimum, temperatures of operation, existing electrotransport and diffusion data for interstitial solutes in solid metals yield values of U/D ranging approximately from 3 to 60. This suggests that electrotransport experiments conducted under optimum conditions should be effective in removing interstitial impurities from at least some metals. Thus, for example, reasonably fast, extensive purification is indicated for the systems Pd-H, Th-N, Th-C, Fe-C, Ta-C, Co-C, Ti-C and Y-O; on the other hand only moderate purification seems likely in the case of systems such as Fe-H, Fe-D, Ni-H, Ni-D, Ta-H, Ti-O, Ni-C and W-C. A similar examination of electrotransport and diffusion data available for substitutional alloys gives values of U/D ranging from 100 - 1000 and extremely large purifications are thus theoretically possible. However, owing to the very low values of the diffusion coefficients for substitutional solutes, the time required to achieve significant removal of such substitutional impurities would in general be prohibitively long. Electrotransport does not appear, therefore, to hold great promise for the separation of substitutional solutes from solids.

The technique has been successfully applied by Williams and Huffine[2] to the purification of solid yttrium. Under the influence of d.c. alone, oxygen and several substitutional impurities (Fe, Mn, Ni, Co, Ti, B) were found to migrate to the anode. The oxygen content at the cathode was reduced by 80% and the metal in this region was found to be ductile; previously, more normal methods of purification had been unable to produce ductile yttrium. Other workers have also demonstrated the effectiveness of electrotransport in removing interstitial impurities (C, N, O) from yttrium[3] and also from vanadium[63], lutetium[79] and thorium[4]. The feasibility of purifying cerium, uranium and plutonium by this means has also been investigated[64]. A review of the experimental factors involved in achieving a high degree of purification of metals by electrotransport may be found in a recent paper by Peterson[109].

In liquid metal systems, purification solely by electrotransport has so far met with little success, mainly because, at high temperatures, melt contamination and stirring arising from thermal and electrokinetic or electro-magnetic convection present particular problems. Using the results of Belaschenko's studies of electrotransport in liquid alloys, Verhoeven[5] has estimated values of U/D between about 2 and 400. For the majority of the systems for which appropriate data are available U/D is greater than 12, indicating that electrotransport could be effective in removing many different impurities from liquid metals. To achieve this in an acceptable time, however, would appear to require field intensities of the order of at least 0.1 volt per cm. Useful application of the technique as a means of purification will therefore depend on the ability to prevent the current-induced convection which is probable at such high field intensities. This should be achievable using suitably designed capillaries; in recent studies of liquid In-Sb alloys[65] convection was successfully suppressed and rapid separation of the components achieved, even with field intensities of 0.5 - 0.7 volt per cm, by the use of ribbon-type capillaries.

Electrotransport purification is inherently an extremely inefficient process, since only a small proportion of the electrical energy consumed is associated with the movement of the atoms and thus effecting purification. Nevertheless there is good evidence for the feasibility of the technique, at least in the case of interstitial solid solutions and the liquid state, and for certain special requirements the high costs might be justified.

Although, principally because of its low efficiency, the value of electrotransport as a primary purification method appears to be limited, it may play a useful role as an auxiliary factor in zone-melting and zone-refining techniques. Its relevance in these fields has been considered by several groups of workers[66-75].

If diffusion away from a freezing interface can not keep pace with the rate of rejection of solute into the liquid, the solute concentration in the liquid at the interface rises. Governed by the effective partition coefficient, k' - the ratio of solute concentration in the freezing solid to that in the bulk liquid, the solute concentration in the solid also rises and the degree of purification falls. Although solute movement away from the interface can be enhanced by agitating the liquid, there remains a "stagnant" boundary layer where solute movement is still by atomic diffusion; stirring reduces but does not eliminate this layer.

The value of the effective partition coefficient, k', depends upon the conditions of freezing. In practice, in the absence of an electric field, k' lies somewhere between the equilibrium value, k'_o, and unity; at high crystal growth rates k' approaches unity, while for very slow growth rates equilibrium values are approached. It has been suggested[66,67] that, by applying a suitable electric field, k' can be made to assume a wider range of values, although there is some dispute about the possibility of k' becoming less than k'_o or greater than unity. Possible applications of electrotransport in the fields of zone-melting and crystal growth may be summarised as follows:

(i) the movement of solute away from an interface may be enhanced so that a build-up of solute is prevented and the value of k' maintained;

(ii) in systems where k'_o is near unity, and zone-refining therefore ineffective, the value of k' might be made to differ sufficiently from unity to make zone-refining worthwhile;

(iii) the value of k' might be adjusted to unity to prevent the separation of alloy components even at very slow growth rates; this effect could assist in controlling composition during the growth of alloy crystals.

Since the feasibility of electric field freezing was first pointed out[66], an increasing amount of experimental data has appeared in the literature[68,69,70,73]. However, it must be remembered that at high current densities only a very small taper is required in the container of a liquid metal to set up magnetohydrodynamic convection[76]. Since such irregularities exist at the solid-liquid interfaces, or at the liquid surface, it has been suggested[69] that only specimens with radii of less than 1-2 mm will experience enhanced purification.

5. References

1. M. Gerardin, Compt. Rend., 1861, 53, 727.
2. J.M. Williams and C.L. Huffine, Nuc. Sci. + Eng., 1961, 9, 500.
3. O.N. Carlson, F.A. Schmidt and D.T. Peterson, J. Less-common Metals, 1966, 10, 1.
4. D.T. Peterson, F.A. Schmidt and J.D. Verhoeven, Trans. Met. Soc. A.I.M.E., 1966, 236, 1311.
5. J.D. Verhoeven, J. Metals, 1966 (Jan.), 26.
6. I.A. Blech and E.S. Meieran, Appl. Phys. Lett., 1967, 11, 263.
7. I.A. Blech and E.S. Meieran, J. Appl. Phys., 1969, 40, 485.
8. P.B. Ghate, Appl. Phys. Lett., 1967, 11, 14.
9. K.E. Schwarz, "Elektrolytische Wanderung in flüssigen und festen Metallen" (J.A. Barth) Leipzig, 1940.
10. W. Jost, "Diffusion in Solids, Liquids and Gases", Academic Press, New York, 1960.
11. W. Seith, "Diffusion in Metallen: Platzwechselreaktionen", 2nd ed. Springer Verlag, Berlin, 1955.
12. Th. Heumann, N.P.L. Symposium "The Physical Chemistry of Metallic Solutions and Intermetallic Compounds" Paper 2C H.M.S.O., London 1959.
13. J.D. Verhoeven, Met. Revs., 1963, 8, 311.
14. D.K. Belaschenko, Russian Chem. Revs., 1965, 34, 219.
15. M.D. Glinchuk, Ukrain, Fiz. Zhur., 1959, 4, 684.
16. F. Skaupy, "Elektrizitätsleitung in Metallen", Verh. deut. phys. Gesellschaft, XVI Jahrgang, 1914, 159.
17. G.N. Lewis, E.Q. Adams and E.H. Lamman, J. Amer. Chem. Soc. 1915, 37, 2656.
18. C. Wagner, Z.phys. Chem., 1932, 15B, 347; 1933, 164A, 231.
19. K. Schwarz, ibid., 1933, 164A, 223.
20. S.I. Drakin, Zhur. Fiz. Khim., 1953, 27, 1586.
21. S.R. De Groot, "Thermodynamics of Irreversible Processes", North-Holland Publishing Co., Amsterdam, 1966.
22. B. Baranowski, Roczniki Chem., 1955, 29, 129.
23. B. Baranowski, ibid., 1955, 29, 586.
24. B. Baranowski, ibid., 1956, 30, 841.
25. B. Baranowski, Bull. Acad. Polon. Sci. Ser. Sci. Math. Astron., 1955, 3, 117.
26. B. Baranowski, ibid., 1956, 4, 465.
27. B. Baranowski and A.S. Cukrowski, ibid., 1962, 10, 135.

28. B. Baranowski and A.S. Cukrowski, Zhur. Fiz. Khim., 1962, 36, 2096. Russ. J. Phys. Chem. 1962, 36, 1129.

29. D.K. Belashchenko and A.A. Zhukhovitskii, Zhur. Fiz. Khim., 1961, 35, 1921, Russ. J. Phys. Chem. 1961, 35, 944.

30. D.K. Belashchenko and B.S. Bokshtein, ibid., 1961, 35, 2228, ibid., 1961, 35, 1099.

31. A. Klemm, Z. Naturforsch., 1953, A, 8, 397.

32. A. Klemm, ibid., 1954, A, 9, 1031.

33. P.C. Mangelsdorf, J. Chem. Physics, 1960, 33, 1151.

34. P.G. de Gennes, J. Phys. Radium, 1956, 17, 343.

35. V.B. Fiks, Fizika Tverdogo Tela, 1959, 1, 16, Soviet Physics - Solid State, 1959, 1, 14.

36. N.F. Mott and H. Jones, "Properties of Metals and Alloys", Oxford, 1936.

37. D.K.Belashchenko, Izv. Vys. Ucheb. Zaved., Chernaya Met., 1961, No.9, 5.

38. D.K.Belashchenko, Izv. Akad. Nauk. S.S.S.R. (Otdel. Tekhn.), Met.i Topl., 1960, (6), 89.

39. D.K.Belashchenko, Zhur, Fiz. Khim., 1961, 35, 1871, Russ. J. Phys. Chem. 1961, 35, 923.

40. D.K.Belashchenko and G.A. Grigor'ev, Izv. Vys. Ucheb. Zaved. Chernaya Met., 1961, (11), 116; 1962 (1), 124; 1962, (5), 120; 1962, (7), 137.

41. S.E. Bresler and G.E. Pikus, Zhur, Tekhn. Fiz., 1958, 28, 2282. Soviet Physics-Technical Phys., 1958, 3, 2094.

42. H.B. Huntington and A.R. Grone, J. Phys. Chem. Solids, 1961, 20, 76.

43. H.B. Huntington and Siu-Chung Ho ,J. Phys. Soc. Japan, 1963, 18 (Suppl. 2), 202.

44. C. Bosvieux and J. Friedel, J. Phys. Chem. Solids, 1962, 23, 123.

45. H.B. Huntington, Trans. Met. Soc. A.I.M.E., 1969, 245, 2571.

46. Yu. G. Miller, Soviet Physics - Solid State, 1962, 3, 1728.

47. M.J. Bibby and W.V. Youdelis, Can. J. Phys., 1966, 44, 2363.

48. Th. Hehenkamp, Habilitationsshrift, Münster, 1966.

49. H.B. Huntington, W.B. Alexander, M.D. Feit and J.L. Routbort, Proc. Europhysics Conf. "Atomic Transport in Solids and Liquids", Marstrand, 1970. p. 91. (Verlag. Z. für Naturforsch., Tubingen, 1971).

50. J.L. Routbort, Phys. Rev. 1968, 176, 796.

51. W.B. Alexander, Z. für Naturforsch., 1971, 26a, 18.

52. S. Brown and S.J. Barnett, Phys. Rev., 1952, 87, 601.

53. A. Lodding, Gothenburg Stud. Phys., 1961, 1.

54. R.A. Oriani and O.D. Gonzalez, Trans. Met. Soc. A.I.M.E., 1967, 239, 1041.

55. W. Seith and H. Wever, Z. für Elektrochemie, 1953, 57, 891.

56. R. Kremann, Monatsh. für Chemie, 1926, 47, 295; 1930, 56, 16, 35; 1931, 57, 323; 1929, 53/54, 203.

57. J. Angus, J.D. Verhoeven and E.E. Hucke, Trans. A.I.M.E., 1959, 7, 447.

58. S.G. Epstein and A. Paskin, Phys. Letters, 1967, 24A, 309.

59. T.E. Faber and J.M. Ziman, Phil. Mag. 1965, 11, 153.

60. R.G.R. Sellors and J.N. Pratt, Phys. and Chem. of Liquids, 1970, 2, 19.

61. J.M. Ziman, Private communication, 1968. (Also refs. 58 and 62).

62. S.G. Epstein, Advances in Physics, 1967, 16, 325.

63. F.A. Schmidt and J.C. Warner, J. Less-common Metals, 1967, 13, 493.

64. R.H. Moore, F.M. Smith and J.R. Morrey, Trans. Met. Soc. A.I.M.E., 1965, 233, 1259.

65. R.G.R. Sellors and J.N. Pratt, Final Techn. Rep. (Min. Tech. Agreement PD/27/024). U. of Birm. 1968.

66. J. Angus, D.V. Ragone and E.E. Hucke, Metallurgical Soc. Conferences 8, 833, Interscience, New York, 1961.

67. W.G. Pfann and R.S. Wagner, Trans. Met. Soc. A.I.M.E., 1962, 224, 1139.

68. D.R. Hay and E. Scala, ibid, 1965, 233, 1153.

69. J.D. Verhoeven, ibid., 233, 1156; 1967, 239, 694; 1967, 239, 761.

70. R.S. Wagner, C.E. Miller and H. Brown, ibid., 1966, 236, 555.

71. D.T.J. Hurle, J.B. Mullin and E.R. Pike, Phil. Mag., 1964, 9, 423.

72. D.T.J. Hurle, J.B. Mullin and E.R. Pike, J. Materials Sci., 1967, 2, 46.

73. F. Claisse, Canad. Met. Quarterly, 1965, 4, 113.

74. I.N. Larionov, N.M. Roizin and V.M. Nogin, Soviet Physics - Semiconductors, 1967, 1, 1175.

75. W.G. Pfann, "Zone Melting", 2nd ed., J. Wiley, New York, 1966.

76. A. Lodding and A. Klemm, Z. Naturforschung, 1962, 17a, 1085.

77. M. Gerl, Z. Naturforsch., 1971, 26a, 1.

78. N.V. Doan, J. Phys. Chem. Solids, 1970, 31, 2079.

79. D.T. Peterson and F.A. Schmidt, J. Less-Common Metals, 1969, 18, 111.

80. S.G. Epstein and J.M. Dickey, Phys. Rev., 1970, B1, 2442.

81. A. Meyer, C.W. Nestor Jr. and W.H. Young, Proc. Phys. Soc., 1967, 92, 446.

82. D.L. Olson, J.L. Blough and D.A. Rigney, Scripta Met., 1970, 4, 1023.

83. D.L. Olson, J.L. Blough and D.A. Rigney, Acta Met., 1972, 20, 305.

84. N. Van Doan, J. Phys. Chem. Solids, 1971, 32, 2135.

85. N. Van Doan and G. Brebec, J. Phys. Chem. Solids, 1970, 31, 475.

86. H.B. Huntington, Encyclopaedia of Chemical Technology p. 278 (J. Wiley N.Y. 1971).

87. Y. Adda and J. Philibert "La Diffusion dans les Solides" Tome II, Presses Universitaire de France, 1966; p. 893.

88. B. Baranowski and A.S. Cukrowski, Archiwum Hutnictwa, 1964, 9, 31.

89. H.J. Stepper and H. Wever, J. Phys. Chem. Solids, 1967, 28, 1103.

90. H. Feller, T. Heumann and H. Wever, Z. für Naturforsch., 1958, 13A, 152.

91. H.M. Gilder and D. Lazarus, Phys. Rev., 1966, 145, 507.

92. N.K., Archipova and S.M. Klotsman, Phys. Stat. Solid, 1966, 16, 729.

93. C.S. Fuller and J.C. Sieverens, Phys. Rev., 1954, 96, 21.

94. R.V. Penney, J. Phys. Chem. Solids, 1964, 25, 335.

95. H. Wever, N.P.L. Symposium "The Physical Chemistry of Metallic Solutions and Intermetallic Compounds", Paper 2L, H.M.S.O., London 1959.

96. S.D. Gertsriken, I. Ya. Dekhtar, V.S. Mikhalenkov and V.M. Falchenko, Ukrain. Fiz. Zhur., 1961, 6, 129.

97. S.G. Epstein, Trans. A.I.M.E., 1966, 236, 1123.

98. S.I. Drakin and A.K. Maltsev, Zhur. Fiz. Khim., 1957, 31, 2036.

99. S.I. Drakin and Yu. K. Titova, Russian J. Phys. Chem., 1967, 41, 319.

100. G.E. Pikus and V.B. Fiks, Fiz. Tverd. Tela., 1959, 1, 1062.

101. P.C. Mangelsdorf, J. Chem. Phys., 1959, 30, 1170.

102. K. Schwarz, Z. für Elektrochemie, 1938, 44, 648.

103. T. Hehenkamp, J. Appl. Phys., 1968, 39, 3928.

104. V.B. Fiks, Proc. Europhys. Conf. "Atomic Transport in Solids and Liquids", Marstrand, 1970, p.3 (Verlag. Z. für Naturforsch., Tubingen, 1971).

105. D.K. Belashchenko, ibid., 1970, p. 173.

106. M. Gerl, ibid., 1970, p. 9.

107. N. Van Doan, ibid., 1970, p. 55.

108. G. Frohberg, ibid., 1970, p. 19.

109. D.T. Peterson, ibid., 1970, p. 104.

110. N. Van Doan, J. Phys. Chem. Solids, 1972, 33, 2161.

111. N. Van Doan and G. Brebec, J. Phys. Chem. Solids, 1973, to be published.

112. J.R. Manning, "Diffusion Kinetics for Atoms in Crystals", Van Nostrand, N.Y., 1968.

113. G.L. Hofman and A.G. Guy, J. Phys. Chem. Solids, 1972, 33, 2167.

114. O.P. Sinha, Physics Letters, 1972, 38A, 193.

115. R. Rosenberg and L. Berenbaum, Proc. Europhys. Conf. "Atomic Transport in Solids and Liquids", Marstrand, 1970, p. 113 (Verlag. Z. für Naturforsch., Tubingen, 1971).

116. P.S. Ho and L.D. Glowinski, ibid., 1970, p. 123.

117. R.E. Hummel and H.M. Breitling, ibid., 1970, p. 127.

118. P. Adam, ibid., 1970, p. 133.

119. F.M. D'Heurle, Met. Trans., 1971, 2, 683.

ELECTROTRANSPORT DATA FOR METALS AND ALLOYS

Section A: Electrotransport in Liquid and Solid Binary Alloys, and Self-electrotransport in Solid Pure Metals.

Section B: Electrotransport in Ternary Alloys.

Section C: Isotope Separation in Pure Metals by Electrotransport, (Haeffner Effect).

Section D: Electrotransport of Solid Intermetallic Compounds in Liquid Metals.

N.B. (1) The references quoted in column 9 appear in a separate list at the end of the table.

(2) For some alloy systems of type XY (Section A), data are reported with solute X in solvent Y and elsewhere in the table with solute Y in solvent X.

(3) Where values for t, U and Z^o (the transference number, the electric mobility, and the effective valence respectively) are summarised in Section A of the table, they are presented in numerical order rather than in order of temperature dependence. In general however, t and U increase, while Z^o decreases with increasing temperature.

(4) In some Russian work[13,18,96,101] data are reported not as t, U or Z^o, but as an electrodiffusion coefficient, K[96], defined as:

$$K = - T/\Delta\Phi_x \,.\, \ln \frac{c_x}{c_o}$$ (Drakin and Titova[96]),

where: c_x = concentration at co-ordinate x.
c_o = concentration at reservoir.
$\Delta\Phi_x$ = potential drop between co-ordinates O and X.
T = temperature.

(5) By convention, a solute accumulating at the positive electrode is regarded as having a negative effective valence, Z^o. This convention is adhered to in column 8 of the table. No sign has been ascribed to the values for U, t or K in the table, although the general convention is to regard these terms as positive for solutes accumulating at the positive electrode.

N.B. (6) Although data for t and U are sometimes reported in absolute terms for both components of a <u>solid</u> alloy, they can only be reported as a differential mobility between the components of a liquid alloy.

Section A

Electrotransport in Liquid and Solid Binary Alloys and Self-Electrotransport in Pure Solid Metals

Group IA Solvents

SOLVENT Li

Solute	Atomic % solute	Temp. (^{o}C)	State	Solute accumulates at:	Transference no. (gm ion $Faraday^{-1}$)	Electric mobility $cm^2v^{-1}sec^{-1}$	Effective valence (Z^o)	Reference
Li	Self ET	96 to 146	solid	A	-	-	-1.71 to -1.33	220

SOLVENT Na

Solute	Atomic % solute	Temp. ($^oC.$)	State	Solute accumulates at:	Transference no. (gm ion Faraday^{-1})	Electric mobility $cm^2 v^{-1} sec^{-1}$	Effective valence (Z^o)	Reference
Hg	0.6	110	liquid	A	-	-	-18	3
Hg	1.1	110	"	A	-	-	7.7	3
Hg	2.89 wt.%	115	"	A	9.3×10^{-6}	-	-	17
Hg	3.0 wt.%	115	"	A	6.0×10^{-6}	-	-	"
Hg	2.99 wt.%	115	"	A	6.2×10^{-6}	-	-	"
Hg	2.99 wt.%	115	"	A	6.0×10^{-6}	-	-	"
Hg	3.09 wt.%	130	"	A	6.0×10^{-6}	-	-	"
Hg	2.21 wt.%	130	"	A	4.5×10^{-6}	-	-	"
Hg	20.95 wt.%	115	"	A	64.7×10^{-6}	-	-	"
Hg	20.0 wt.%	115	"	A	64.0×10^{-6}	-	-	"
Hg	16.75 wt.%	115	"	A	38.6×10^{-6}	-	-	"
Hg	19.70 wt.%	215	"	A	69.0×10^{-6}	-	-	"
Hg	19.45 wt.%	215	"	A	56.2×10^{-6}	-	-	"
Hg	20.5 wt.%	215	"	A	59.4×10^{-6}	-	-	"
Hg	0.39 to 0.45	154	"	A	-	420×10^{-4}	-50	163
Hg	2.11 wt.%	140	"	A	-	$K=(3.83 \text{ to } 4.35) \times 10^{+5}$	-	96
Hg	3.44 wt.%	140	"	A	-	$K=(4.21 \text{ to } 4.85) \times 10^{+5}$	-	"
Hg	6.60 wt.%	140	"	A	-	$K=(3.80 \text{ to } 4.45) \times 10^{+5}$	-	"

SOLVENT Na (Continued)

Solute	Atomic % solute	Temp. (°C.)	State	Solute accumulates at:	Transference no. (gm ion Faraday^{-1})	Electric mobility $cm^2v^{-1}sec^{-1}$	Effective valence (Z^o)	Reference
Hg	9.25 wt.%	140	liquid	A	-	K=(3.18 to 3.74) x 10^{+5}	-	96
Cd	dilute	150 to 300	"	A	-	2.5 x 10^{-2}	-	94
Cd	0.21	110	"	A	-	-	-13, -7.3	3,13,16
						(K = 1.7 x 10^{+5}, Z^o	corrected to -15)	96
Cd	10^{-3} to 1.0	130 to 530	"	A	-	-	-80 to -6	219
Cd	0.7 to 1.6	150 to 330	"	A	-	250 x 10^{-4}	-	162
Cd	0.34 to 0.67	155	"	A	-	400 x 10^{-4}	-43	163
Sn	0.09 to 0.1	200	"	A	-	460 x 10^{-4}	-27 -21, -27.2	163 3,16
Pb	0.45	115	"	A	-	(K = 6.3 x 10^{+5}, Z^o	corrected to -54)	96
K	1.6	100	"	A	-	-	0.84	4,12
K	50	100	"	C	-	-	-	1
K	3 to 97 wt.%	100	"	A, C(a)	-	-	-	4
K	30	85 to 300	"	A(b)	-	-	Z^o_{Na} = +0.009 Z^o_K = -0.023	184
K	47	85 to 300	"	C(b)	-	-	Z^o_{Na} = - 0.13 Z^o_K = +0.015	184
K	1.1 to 1.3	108 to 304	"	A	-	-	-	190

SOLVENT Na (Continued)

Solute	Atomic % solute	Temp. (°C.)	State	Solute accumulates at:	Transference no. (gm ion Faraday^{-1})	Electric mobility $cm^2v^{-1}sec^{-1}$	Effective valence (Z^o)	Reference
K	9.8 to 10.2	95 to 213	liquid	A	-	-	-	190
K	32.8 to 35.5	51 to 290	"	null point (c)	-	-	-	"
K		42	"	A (K < 36%) C (K > 36%)	-	-	-	199
K		167	"	A (K < 31%) C (K > 31%)	-	-	-	199
Rb	1.0 to 95	81 to 137	"	A (Rb < 30%) C (Rb > 30%)	-	-	-	198
In	0.4 to 1.1	146 to 300	"	A	-	400×10^{-4}	-	162
In	3 wt.%	115	"	A(d)	-	(K=$1.9 \times 10^{+5}$) (K=$3.7 \times 10^{+5}$)	- -32	18 96
Ba	5 wt.%	115	"	A(d)	-	(K=$1.1 \times 10^{+5}$) (K=$2.1 \times 10^{+5}$)	- -18	18 96
Ba	8.4 wt.%	115	"	A(d)	-	(K=$0.7 \times 10^{+5}$)	-	18
Ag	2.5 wt.%	150	"	no effect (d)	-	(K less than $0.3 \times 10^{+5}$)	0	18
Sr	3.7 wt.%	115	"	no effect (d)	-	(K less than $0.3 \times 10^{+5}$)	0	18
Sr	8.4 wt.%	115	"	no effect (d)	-	(K less than $0.3 \times 10^{+5}$)	0	18
Tl	5 wt.%	-	"	A(d)	-	(K=$1.7 \times 10^{+5}$)	-	13,18
Tl	4.1 wt.%	115	"	A(d)	-	10.7×10^{-6}	-	17,18
Tl	5.35 wt.%	115	"	A(d)	-	16.9×10^{-6} (K=$3.4 \times 10^{+5}$)	- -29	17,18 96

SOLVENT Na (Continued)

Solute	Atomic % solute	Temp. (°C.)	State	Solute accumulates at:	Transference No. (gm ion $Faraday^{-1}$)	Electric mobility $cm^2v^{-1}sec^{-1}$	Effective valence (Z^o)	Reference
Tl	3.18	115	"	A	-	7.9×10^{-6}	-	17,18
Na	(Self ET)	82.8 to 94.4	Solid	A	-	-	-2.4	119

(a) Electrode at which solute accumulates changes from A to C above 48 wt. % Na.

(b) Electrode at which solute accumulates changes from A to C above 39 at. % K.

(c) Electrode at which solute accumulates changes from A to C above ~34 at. % K.

(d) In, Tl and Ba form compounds with Na, but Ag and Sr do not.

SOLVENT K

Solute	Atomic % solute	Temp. (°C.)	State	Solute accumulates at:	Transference no. (gm ion $Faraday^{-1}$)	Electric mobility $cm^2v^{-1}sec^{-1}$	Effective valence (Z^o)	Reference
Na	5.5	100	liquid	A	-	-	-0.46	4
Na	3 to 97 wt.%	100	"	Ac(a)	-	-	-	4
Na	63.0 to 67.2	51 to 290	"	null point (b)	-	-	-	190
Na	1.42 to 1.46	76 to 258	"	A (b)	-	-	-	190
Na		42	"	A (Na < 64%) C (Na > 64%)	-	-	-	199
Na		167	"	A (Na < 69%) C (Na > 69%)	-	-	-	199
Hg	0.4	100	"	A	-	-	-10,-12.6	6,16
Hg	2.42 wt.%	120	"	A	12.4×10^{-6}	-	-	17
Hg	2.05 wt.%	120	"	A	7.7×10^{-6}	-	-	17
Hg	2.88 wt.%	120	"	A	11.6×10^{-6}	-	-	17
Hg	1.90 wt.%	120	"	A	9.8×10^{-6}	-	-	17
Hg	3.15 wt.%	120	"	A	16.2×10^{-6}	-	-	17
Hg	3.07 wt.%	120	"	A	13.4×10^{-6}	-	-	17
Hg	10.80 wt.%	115	"	A	43.2×10^{-6}	-	-	17
Hg	11.11 wt.%	115	"	A	46.2×10^{-6}	-	-	17
Hg	16.38 wt.%	115	"	A	51.8×10^{-6}	-	-	17
Hg	16.94 wt.%	115	"	A	57.8×10^{-6}	-	-	17

SOLVENT K (Continued)

Solute	Atomic % solute	Temp. (°C.)	State	Solute accumulates at:	Transference no. (gm ion $Faraday^{-1}$)	Electric mobility $cm^2v^{-1}sec^{-1}$	Effective valence (Z^o)	Reference
Tl	0.19	110	liquid	A	-	-	-20	3
						(K=4.7 x 10^{+5}, Z^o corrected to -41)		96
Tl	1.65 wt.%	115	liquid	A	-	41 x 10^{-6}	-	17
Tl	1.81 wt.%	115	"	A	-	84 x 10^{-6}	-	17
Tl	1.97 wt.%	115	"	A	-	62 x 10^{-6}	-	17
Tl	1.85 wt.%	115	"	A	-	86 x 10^{-6}	-	17
Pb	0.6	110	"	A	-	-	-22	6
						(K=5.1 x 10^{+5}, Z^o corrected to -44)		96

(a) Electrode at which solute accumulates changes from A to C above 48 wt.% Na.

(b) Electrode at which solute accumulates changes from A to C above ~66 at.% Na.

Group IB Solvents

SOLVENT Cu

Solute	Atomic % solute	Temp. (^{o}C.)	State	Solute accumulates at:	Transference no. (gm ion Faraday^{-1})	Electric mobility cm^{2}v^{-1}sec^{-1}	Effective valence (Z^{o})	Reference
H	-	-	liquid	C	-	-	-	11
Au	38	1000	solid	A	for Cu 7.4 x 10^{-11}	-	-	78
Ag	trace	789 to 1062	"	A	-	-	-2 to -32.3	164
Al	20.8 to 26.1	890 to 905	"	A (and Cu to A)	for Cu 46 to 79 x 10^{-8}	for Al 4.2 x 10^{-6}	-	22,27
					for Al 2.4 to 3.5 x 10^{-8}	for Cu 4.4 x 10^{-6}	-	
Al	50 wt.%	1050	liquid	C	-	-	-	1
Al	2(α)	950	solid	A	for (Al+Cu)5x10^{-9}	-	-	22,36
Al	24(β)	895	"	A	for (Al+Cu)2.8x10^{-6}	-	-	22,37,89
Al	24(β)	895	"	A	for (Al+Cu)7.5x10^{-7}	-	-	22,37,89
Al	24(β)	895	"	A	for (Al+Cu)4.5x10^{-7}	-	-	22,37,89
Al	33(γ)	900	"	C	for (Al+Cu)2 x10^{-6}	-	-	22,37,89
Sn	trace	1075 to 1153	liquid	A	-	-	-57.0 to -19.9	60
Sn	0.1	1101 to 1159	"	A	-	-	-51.0 to -27.6	60
Sn	20(γ)	670	solid	A	for (Sn+Cu)8.5x10^{-7}	-	-	22,36
Sn	20(γ)	610	"	A	for (Sn+Cu)4.5x10^{-7}	-	-	22,36
Sn	20(δ)	575	"	C	for (Sn+Cu)1.5x10^{-7}	-	-	22,36
Sn	24(ϵ)	530	"	A	for (Sn+Cu)5 x10^{-8}	-	-	22,36

SOLVENT Cu (Continued)

Solute	Atomic % solute	Temp. (^oC.)	State	Solute accumulates at:	Transference no. (gm ion Faraday^{-1})	Electric mobility $cm^2 v^{-1} sec^{-1}$	Effective valence (Z^o)	Reference
Sn	44.5(η)	235	solid	A	for Sn+Cu 3×10^{-8}	-	-	22,36
							exp, theory	
Sn	0.8 to 1.0	735	"	A	-	-	-246, -14.8	134
Sn	0.8 to 1.0	759	"	A	-	-	-425, -14.5	134
Sn	0.8 to 1.0	785	"	A	-	-	-418, -14.1	134
Sn	0.8 to 1.0	836	"	A	-	-	-414, -13.3	134
Sn	0.8 to 1.0	864	"	A	-	-	-302, -13.0	134
Ni	0.8 to 1.0	633	"	A	-	-	$-1.6 \times 10^{+5}$(a) -7.5	134
Ni	0.8 to 1.0	733	"	A	-	-	$-1.6 \times 10^{+4}$(a), -6.5	134
Ni	0.8 to 1.0	855	"	A	-	-	$-2.7 \times 10^{+3}$(a), -5.8	134
Sb	0.8 to 1.0	581	"	A	-	-	-318, -22.0	134
Sb	0.8 to 1.0	595	"	A	-	-	-564, -21.7	134
Sb	0.8 to 1.0	617	"	A	-	-	-398, -7.5	134
Sb	0 to 3.0	718 to 947	"	A	-	-	-104 to -24	217
Sb	trace	790 to 870	"	A	-	-	-33.6 to -49.0	64
Co	trace	882 to 956	"	A	-	-	-26.6 to -38.7	64

SOLVENT Cu (Continued)

Solute	Atomic % solute	Temp. ($^oC.$)	State	Solute accumulates at:	Transference no. (gm ion $Faraday^{-1}$)	Electric mobility $cm^2v^{-1}sec^{-1}$	Effective valence (Z^o)	Reference
Fe	trace	1036 to 1053	solid	A	-	-	-27.5 to -65.7	64
Fe	trace	994	"	A	-	-	-51.74[e](±4.4)	206
Si	dilute	-	liquid	C	-	-	-	185
Ge	dilute	-	"	A	-	-	-	185

SOLVENT Cu (continued)

Solute	Atomic % solute	Temp. (°C.)	State	Solute accumulates at:	Transference no. (gm ion Faraday^{-1})	Electric mobility $cm^2v^{-1}sec^{-1}$	Effective valence (Z^o)	Reference
Cu (selfET)	-	800	solid	A	-	-	-5	116
Cu "	-	900	"	A	-	-	-2	116
Cu "	-	1000	"	no effect (b)	-	-	0	116
Cu "	-	1040	"	C	-	-	-	85,116
Cu "	-	900	"	A	3×10^{-9}	-	-14	90
Cu "	-	1000	"	C(c)	3×10^{-9}	(reversal)	+12	90
Cu (selfET)	-	900	"	A	-	-	-38	92,102
Cu "	-	950	"	A	18×10^{-9}	-	-18	92,102
Cu "	-	1000	"	A	56×10^{-9}	-	-22	92,102
Cu "	-	1004 to 1030	"	A(d)	-	-	-5.5	117
Cu "	(Displacement of grain boundaries towards A)				-	-	-	120
Cu "	-	800 to 1030	"	A	-	-	-8(±3)	216

(a) Electrotransport of Ni in Cu may be a dislocation pipe mechanism.

(b) Direction of migration changes from A to C above 1000° C.

(c) Direction of migration changes from A to C above 900° C.

(d) Migration direction has not reversed; suggests previously observed reversals due to the presence of impurities.

(e) Apparent effective valency (Z_i^{oo}).

SOLVENT Ag

Solute	Atomic % solute	Temp. (°C.)	State	Solute accumulates at:	Transference no. (gm ion Faraday^{-1})	Electric mobility $cm^2 v^{-1} sec^{-1}$	Effective valence (Z^o)	Reference
Zn	30	500 to 560	solid	A	-	for Zn 1.5×10^{-6} for Ag 3×10^{-6}	-	22
Zn	32(α)	550	"	A	for Zn + Ag 3×10^{-8}	-	-	22
Zn	32(α)	550	"	A	for Zn + Ag 1×10^{-7}	-	-	22
Zn	48(β)	500	"	A	for Zn + Ag 8.3×10^{-7}	-	-	22
Zn	48(β)	500	"	A	for Zn + Ag 4.6×10^{-7}	-	-	22
Zn	48(β)	500	"	A	for Zn + Ag 6×10^{-7}	-	-	22
Zn	48(β)	335	"	A	for Zn + Ag 8×10^{-8}	-	-	22
Zn	62(γ)	500	"	C	for Zn + Ag 3×10^{-7}	-	-	22
Zn	71(ε)	500	"	A	for Zn + Ag 7×10^{-7}	-	-	22
Zn	25	550	"	Both to A	for Zn 0.97×10^{-7}(a) for Ag 8.71×10^{-7}	-	-	74
Zn	35	550	"	Both to A	for Zn 2.67×10^{-7}(a) for Ag 7.16×10^{-7}	-	-	74
Zn	50	550	"	Both to A	for Zn 5.73×10^{-7}(a) for Ag 1.35×10^{-7}	-	-	74
Zn	50	434	"	Both to A	-	-	-0.6	62
Zn	30	500 to 600	"	Ag to A	-	3×10^{-6}	-	22
Si	dilute	-	liquid	A	-	-	-	185
Ge	dilute	-	"	A	-	-	-	185

SOLVENT Ag (Continued)

Solute	Atomic % solute	Temp. (°C.)	State	Solute accumulates at:	Transference no. (gm ion Faraday^{-1})	Electric mobility $cm^2v^{-1}sec^{-1}$	Effective valence (Z^o)	Reference
Au	10-90	925-940	solid	A(Ag + Au)	-	-	Z^o_{Au}=-9.9 to -8.8 Z^o_{Ag}=-9.4 to -7.7	215
Cd	trace	807-882	"	A	-	-	-22 to -20	207
In	trace	788-880	"	A	-	-	-41.1 to -36.1	207
Sn	trace	703-932	"	A	-	-	-67 to-59.6	207
Sn	1.0	997-1205	liquid	A	-	-	-48.1 to -23.4	60
Sn			solid	A	-	-	-10 to -7	217
Sb	trace	765-933	"	A	-	-	-86.5 to -102	179
Sb			"	A	-	-	-80 to -85	217
Mn	trace	873-877	"	A	-	-	-12.21[b](±2.3) to -9.83[b](±3.2)	206
Fe	trace	880	"	A	-	-	- 52.41[b](±3.8)	206
Co	trace	880-882	"	A	-	-	-38.53[b](±5.2) to -43.32[b](±4.6)	206
Ni	trace	876	"	A	-	-	-20.97[b](±9.2)	206
Po	trace	350	"	A	-	-	-	1
O	40 ppm	700-900	"	A	-	-	-6.5 ± 1.4 N.B.	222

SOLVENT Ag (Continued)

Solute	Atomic % solute	Temp. ($^oC.$)	State	Solute accumulates at:	Transference no. (gm ion $Faraday^{-1}$)	Electric mobility $cm^2v^{-1}sec^{-1}$	Effective valence (Z^o)	Reference
Ag (SelfET)	-	745	solid	A	-	-	-26	92
Ag (SelfET)	-	800	"	A	34×10^{-9}	-	-27	92,102
Ag (SelfET)	-	875	"	A	-	-	-22	92
Ag (SelfET)	-	899	"	A	91×10^{-9}	-	-16.5	92,102
Ag (SelfET)	-	837	"	A	9.6×10^{-9}	-	-7.6	102
Ag (SelfET)	-	899	"	A	42×10^{-9}	-	-7.5	102
Ag (SelfET)	1 ppm O	670 to 880	"	A	-	approx. (3 to 20) $\times 10^{-12}$ $cm^3 \cdot amp^{-1} \cdot sec^{-1}$	-25(mean) -10.0±0.5 (corrected)	23,108 222
Ag (SelfET)	-	600 to 900	"	A	-	-	-17 to -24	124
Ag (SelfET)	-	765 to 930	"	A	-	-	-27.7 to -16.1	179
Ag (SelfET)	40 ppm O	700 to 900	"	A	-	-	-6.5 ± 1.4	222

[a] i. e. Ratio of rates of ion transport reverses with increasing composition. [b] Apparent effective valencies (Z_i^{oo}).

SOLVENT Au

Solute	Atomic % solute	Temp. (°C)	State	Solute accumulates at:	Transference no. (gm ion $Faraday^{-1}$)	Electric mobility $cm^2 v^{-1} sec^{-1}$	Effective valence (Z^o)	Reference
In	0.3	800 to 950	solid	A	-	-	-9.3 to -7.4	214
Sn	0.3	800 to 950	"	A	-	-	-19.4 to -43.1	214
Si	dilute	-	liquid	C	-	-	-	185
Ge	27	487	"	C	-	-	-	106
Ge	dilute	-	"	C	-	-	-	185
Sb	trace	853, 1009	solid	A	-	9.24×10^{-11}, 4.54×10^{-10} cm^3/amp. sec.	-140(±40)	111
Sb	0.3	800 to 950	"	A	-	-	-43.1 to -34.5	214
Pd	29	900	"	C	1.6×10^{-4}	-	-	79,81
Cu	35	750	"	C	-	-	-	79,81
Ag	trace	797 to 922	"	A	-	-	-1.9 to -13.1	164
Ag	10-90	925 to 940	"	A (Ag + Au)	-	-	Z^o_{Ag} = -7.7 to -9.4 Z^o_{Au} = -8.8 to -9.9	215
Ag	50	860 to 940	"	A (Ag + Au)	-	-	Z^o_{Ag} = -8 to -8.9	215

SOLVENT Au (continued)

Solute	Atomic % solute	Temp. (^{o}C.)	State	Solute accumulates at:	Transference no. (gm ion Faraday^{-1})	Electric mobility $cm^2 v^{-1} sec^{-1}$	Effective valence (Z^o)	Reference
Au (SelfET)	-	800	solid	A	-	-	-8.3	88
Au (SelfET)	-	929	"	A	$(16.4 + 90) \times 10^{-9}$	-	-6.6 + -32	88, 92, 102
Au (SelfET)	-	1000	"	A	32.8×10^{-9}	-	-5.5	88, 92, 102
Au (SelfET)	-	750	"	A	-	-	-38	92
Au (SelfET)	-	800	"	A	-	-	-34	92, 109
Au (SelfET)	-	862	"	A	-	-	-	193, 194
Au (SelfET)	-	(Displacement of grain boundaries towards C)				-	-	120
Au (SelfET)	-	450	solid	A	-	-	-	174
Au (SelfET)	-	874-1016	"	A	-	9.6×10^{-13} to 7.3×10^{-12} $cm^3 amp^{-1} sec^{-1}$	-9.0 (±1)	110

Group II Solvents

SOLVENT Zn

Solute	Atomic % solute	Temp. (^{o}C.)	State	Solute accumulates at:	Transference no. (gm ion $Faraday^{-1}$)	Electric mobility $cm^2 v^{-1} sec^{-1}$	Effective valence (Z^{o})	Reference
Ag	50	670	solid	A	-	-	-1.6	62
Ag	50	519	"	A	-	-	-1.9	62
Ag	30	577	"	A	-	-	-33	62
Ag	0.0031	520	liquid	A	-	-	-1.13	186
Ag	3.46	520	"	A	-	-	-1.12	186
Ag	0.0034	560	"	A	-	-	-1.1	186
Ag	3.5	560	"	A	-	-	-1.04	186
Ag	0.0032	620	"	A	-	-	-1.0	186
Ag	0.59	620	"	A	-	-	-1.1	186
Zn (SelfET)	-	350	solid	A	-	-	-4.0	92
Zn (SelfET)	-	370	"	A	-	-	-4.3	92
Zn (SelfET)	-	360-395	"	A	-	-	-2.0(|| to c-axis) -4.4(in basal plane)	208

SOLVENT Cd

Solute	Atomic % solute	Temp. (°C.)	State	Solute accumulates at:	Transference no. (gm ion Faraday^{-1})	Electric mobility $cm^2 v^{-1} sec^{-1}$	Effective valence (Z^o)	Reference
Co	trace	350	liquid	A	-	-	-22	69,75,76
Ag	trace	350	"	A	-	-	-1.25	7,12
Au	trace	350	"	A	-	-	-2.2	7,12
Bi	2.1	350	"	A	-	-	-16.4	8,9,13
Bi	25 to 75	350	"	A	-	-	-	1
Bi	0 to 80	300	"	A	-	-	-13.12 to -0.31	219
Sn	0.37	360	"	A	-	-	-5.4	8,9
Sn	25 to 75	300	"	A	-	-	-	1
Sn	0 to 80	300	"	A	-	-	-5.4 to -0.17	219
Pb	25 to 75	300	"	A	-	-	-	1
Pb	0 to 80	350	"	A	-	-	-5.94 to -0.27	219
Pb	3 wt.%	360	"	A	-	-	-	1,71,75
Tl	trace	350	"	A	-	-	-2.1	69,75
Ni	trace	350	"	A	-	-	-8.4	69,75,76
Cd (SelfET)	-	250	solid	A	-	-	-5	92
Cd (SelfET)	-	286	"	A	-	-	-1.6(‖ to c-axis) -3.3(in basal plane)	175,209

SOLVENT Hg

Solute	Atomic % solute	Temp. ($^oC.$)	State	Solute accumulates at:	Transference no. (gm ion Faraday^{-1})	Electric mobility $cm^2 v^{-1} sec^{-1}$	Effective valence (Z^o)	Reference
Li	dilute	25	liquid	C	-	114×10^{-6}	+ 0.31	1,2,83
Li	dilute	25 to 75	"	C	-	-	-	165
Li	0.24 + 0.58wt.%	240	"	C	$(2.9 \text{ and } 4.6) \times 10^{-6}$	-	-	72,73
Na	dilute	25 to 75	"	A	-	-	-	165
Na	2.89	-	"	A	-	-	-	70,75
Na	dilute	25	"	A	-	$(119 \text{ to } 123) \times 10^{-6}$	-0.43	1,2,5,75
Na	0-100	40	"	A, C(a)	-(reversal)	-	-	1
Na	0.097 to 0.485 wt.%	232 to 334	"	A, C(b)	-(reversal)	$(150 \text{ to } 180) \times 10^{-6}$	-	5
Na	81.1 to 87.7	240	"	C	$(0.8 \text{ to } 6.5) \times 10^{-5}$	-	-	72,73
Na	20.22	240	"	C	1.28×10^{-5}	-	-	72,73
Na	15.23	240	"	A, C(c)	approx. 0 (reversal)	-	-	72,73
Na	11.89	240	"	A	1.94×10^{-5}	-	-	72,73
Na	8.02	240	"	A	1.23×10^{-5}	-	-	72,73
Na	4.76	240	"	A	0.69×10^{-5}	-	-	72,73
Na	3.24	240	"	A	0.29×10^{-5}	-	-	72,73
Na	0.577	240	"	A	0.29×10^{-5}	-	-	70,72,73
Na	-	-	"	C(unreliable)	-	-	-	93
K	0 to 100	240	"	A, C(d)	-(reversal)	-	-	1

SOLVENT Hg (Continued)

Solute	Atomic % solute	Temp. (°C.)	State	Solute accumulates at:	Transference no. (gm ion Faraday^{-1})	Electric mobility $cm^2 v^{-1} sec^{-1}$	Effective valence (Z^o)	Reference
K	36.05 wt.%	240	liquid	C	37.3×10^{-6}	-	-	72,73
K	3.10	240	"	C	17.5×10^{-6}	-	-	72,73
K	2.09	240	"	A, C(e)	12.7×10^{-6} (reversal)	-	-	72,73
K	1.33	240	"	A	9.5×10^{-6}	-	-	72,73
K	1.18	240	"	A	9.6×10^{-6}	-	-	72,73
K	0.55	24	"	A	6.0×10^{-6}	-	-	72,73
K	dilute	25 to 75	"	C	-	-	-	165
K	dilute	25	"	A	-	361×10^{-6}	-1.31	1,2,75
K	dilute	25	"	A	-	570×10^{-6}	-	2
K	-	-	"	C(unreliable)	-	-	-	93
Ba	dilute	25 to 75	"	A	-	-	-	165
Ba	0 to 100	-	"	A, C(f)	-(reversal)	-	-	1
Ba	3.22 to 6.64	240	"	A, C(g)	$(0.49 \text{ to } 0.65) \times 10^{-5}$ (reversal)	-	-	72,73
Ag	trace	25	"	C	-	611×10^{-6}	+1.43	1,7,83
Au	trace	25	"	C	-	$(417 \text{ and } 432) \times 10^{-6}$	+1.47	1,2,75,83
Au	-	-	"	C(unreliable)	-	-	-	93
Mg	dilute	25	"	C	-	$(917 \text{ and } 972) \times 10^{-6}$	+1 approx.	1,2,75,83
Zn	dilute	25	"	C	-	$(1167 \text{ and } 1250) \times 10^{-6}$	+1.25	1,2,83
Cd	50 wt.%	300	"	C	-	-	-	1

SOLVENT Hg (Continued)

Solute	Atomic % solute	Temp. (°C.)	State	Solute accumulates at:	Transference no. (gm ion Faraday^{-1})	Electric mobility $cm^2 v^{-1} sec^{-1}$	Effective valence (Z^o)	Reference
Cd	0.5 to 3.6	25,27 + 43	liquid	C	-	(1056, 972 + 1000) x 10^{-6}	+1.25	1,2,26,83
Cd	0.4 to 2.5	150	"	C	-	(1190 to 1140) x 10^{-6}	-	1,2
Ga	dilute	25	"	C	-	(793 + 806) x 10^{-6}	+1 approx.	1,2,83
In	dilute	25	"	C	-	847 x 10^{-6}	-	2
Tl	dilute	25	"	no effect	0	0	0	1,2,83
Sn	dilute	25	"	C	-	(528 + 639) x 10^{-6}	+0.65	1,2,83
Pb	dilute	25	"	no effect	0	0	0	1,2,83
Bi	dilute	25	"	A	-	(806 + 1083) x 10^{-6}	-1.38	1,2,83
Bi	1.65 to 8.84	240	"	A	(0.36 to 2.81) x 10^{-4}	-	-	72,73
Bi	-	-	"	C(unreliable)	-	-	-	93
Ca	-	280	"	C	-	-	-	1,75
Ca	-	25	"	C	-	<56 x 10^{-6}	<+0.1	1,83
Cs	-	25	"	A	-	1194 x 10^{-6}	-4.87	1,83

(a) Electrode at which solute accumulates changes from A to C at 1.5 wt.% Na.

(b) Electrode at which solute accumulates changes from A to C at 290°C.

(c) Electrode at which solute accumulates changes from A to C at 15 at.% Na.

(d) Electrode at which solute accumulates changes from A to C at 2.5 wt.% K.

(e) Electrode at which solute accumulates changes from A to C at 11 at.% K.

(f) Electrode at which solute accumulates changes from A to C at 2.7 wt.% Ba.

(g) Electrode at which solute accumulates changes from A to C at 3.9 at.% Ba.

Group III A Solvents

SOLVENT Y

Solute	Atomic % solute	Temp. ($^{o}C.$)	State	Solute accumulates at:	Transference no. (gm ion $Faraday^{-1}$)	Electric mobility $cm^2 v^{-1} sec^{-1}$	Effective valence (Z^o)	Reference
O	trace	1230 to 1370	solid	A	-	-	-	21
(N)	trace	1230 to 1370	"	(A)	-	-	-	21
Fe	trace	1230 to 1370	"	A	-	-	-	21
Mn	trace	1230 to 1370	"	A	-	-	-	21
Ni	trace	1230 to 1370	"	A	-	-	-	21
B	trace	1230 to 1370	"	A	-	-	-	21
Ti	trace	1230 to 1370	"	A	-	-	-	21
Co	trace	1230 to 1370	"	A	-	-	-	21
(F)	trace	1230 to 1370	"	(A)	-	-	-	21
C	trace	1235	"	A	-	5.56×10^{-5}	-0.5	105
C	(170 ppm)	1350	"	A	-	-	-	105
C		1460	"	A	-	36.11×10^{-5}	-1.1	105
O	trace	1235	"	A	-	8.61×10^{-5}	-1.2	105
O	(780 ppm)	1350	"	A	-	20.28×10^{-5}	-2.6	105
O		1460	"	A	-	31.67×10^{-5}	-1.8	105
N	trace	1235	"	A	-	5.00×10^{-5}	-2.1	105
N	(20 ppm)	1350	"	A	-	10.28×10^{-5}	-2.8	105
N		1460	"	A	-	23.06×10^{-5}	-0.9	105
H	trace	775	"	A	-	6.67×10^{-5}	-0.9	105

SOLVENT Y (Continued)

Solute	Atomic % solute	Temp. ($^oC.$)	State	Solute accumulates at:	Transference no. (gm ion $Faraday^{-1}$)	Electric mobility $cm^2v^{-1}sec^{-1}$	Effective valence (Z^o)	Reference
H	(15 ppm)	835	solid	A	-	10.28×10^{-5}	-0.32	105
H		950	"	A	-	17.50×10^{-5}	-0.26	105

SOLVENT Ce

Solute	Atomic % solute	Temp. ($^oC.$)	State	Solute accumulates at:	Transference no. (gm ion Faraday^{-1})	Electric mobility $cm^2v^{-1}sec^{-1}$	Effective valence (Z^o)	Reference
Fe	trace	495	solid	A	5.6×10^{-6}(a)	-	-	122
Fe	trace	550	"	A	5.6×10^{-6}(a)	-	-	122
Fe	trace	600	"	A	$(4.3 \text{ to } 5.1) \times 10^{-6}$(a)	-	-	122
Fe	trace	650	"	A	5.6×10^{-6}(a)	-	-	122
Co	trace	495 to 650	"	A	-	-	-	122
Ni	trace (35 ppm)	600	"	A	9.9×10^{-8}(a)	-	-	122
Cu	trace	-	"	A	-	-	-	122
Mn	trace	-	"	A	-	-	-	122
O	-	-	"	A	-	-	-	172
Mo	trace	-	"	A	-	-	-	122
C	trace	-	"	no effect	-	-	-	122
Zr	trace	-	"	no effect	-	-	-	122
Sb	trace	-	"	no effect	-	-	-	122
Si	trace	-	"	no effect	-	-	-	122
Mg	trace	-	"	no effect	-	-	-	122

(a) Units are in gm. atoms per sec. per amp. per ohm - cm.

SOLVENT Gd

Solute	Atomic % solute	Temp. (^{o}C.)	State	Solute accumulates at:	Transference no. (gm ion Faraday^{-1})	Electric mobility $cm^2v^{-1}sec^{-1}$	Effective valence (Z^o)	Reference
O	-	-	solid	A	-	-	-	173

SOLVENT Lu

Solute	Atomic % solute	Temp. ($^oC.$)	State	Solute accumulates at:	Transference no (gm ion $Faraday^{-1}$)	Electric mobility $cm^2v^{-1}sec^{-1}$	Effective valence (Z^o)	Reference
C	20 ppm	1330	solid	A	-	2.2×10^{-5}	-2.4	182
C	20 ppm	1450	"	A	-	4.2×10^{-5}	-3.8	182
C	20 ppm	1600	"	A	-	5.5×10^{-5}	-5.1	182
N	20 ppm	1330	"	A	-	6.9×10^{-5}	-3.2	182
N	20 ppm	1450	"	A	-	14×10^{-5}	-3.5	182
N	20 ppm	1600	"	A	-	22×10^{-5}	-5.7	182
O	100 ppm	1330	"	A	-	27×10^{-5}	-2.9	182
O	100 ppm	1450	"	A	-	39×10^{-5}	-2.9	182
O	100 ppm	1600	"	A	-	71×10^{-5}	-2.9	182

SOLVENT Th

Solute	Atomic % solute	Temp. ($^oC.$)	State	Solute accumulates at:	Transference no. (gm ion Faraday^{-1})	Electric mobility $cm^2v^{-1}sec^{-1}$	Effective valence (Z^o)	Reference
C	trace	1440	solid	A	-	1.17×10^{-4}	-2.1	129
C	(50 ppm)	1520	"	A	-	1.47×10^{-4}	-1.9	129
C		1640	"	A	-	3.06×10^{-4}	-2.6	129
C		1675	"	A	-	7.22×10^{-4}	-3.8	129
N	trace	1480	"	A	-	6.94×10^{-4}	-4.4	129
N	(35 ppm)	1520	"	A	-	9.17×10^{-4}	-9.3	129
N		1620	"	A	-	11.67×10^{-4}	-3.2	129
N		1685	"	A	-	12.50×10^{-4}	-2.6	129
O	(trace)	1450	"	A	-	8.33×10^{-4}	-	129
O	(80 ppm)	1540	"	A	-	22.22×10^{-4}	-	129
O		1680	"	A	-	26.94×10^{-4}	-	129
O	-	-	"	A	-	-	-	171

SOLVENT U

Solute	Atomic % solute	Temp. (°C.)	State	Solute accumulates at:	Transference no. (gm ion $Faraday^{-1}$)	Electric mobility $cm^2 v^{-1} sec^{-1}$	Effective valence (Z^o)	Reference
Fe	trace (260 ppm)	900	solid	A	1.7×10^{-8}(a)	-	-	122
Sn	trace	-	"	C	-	-	-	122
O	Oxide UO_2	900 to 1100	"	O to A	$(3.5 \text{ to } 200) \times 10^{-6}$	-	-	131
O	Oxide UO_2	20 to 1000	"	O to A U to C	for U at 1000°C: 3.5×10^{-21}	for U at 1000°C: 7.35×10^{-26} for O at 1000°C: 2.1×10^{-5}	-	132
U (SelfET)	-	880 to 1050	"	A	-	-	-1.2	210

(a) Units are in gm. atoms per sec. per amp. per ohm - cm.

SOLVENT Pu

Solute	Atomic % solute	Temp. (^{o}C.)	State	Solute accumulates at:	Transference no. (gm ion Faraday^{-1})	Electric mobility $cm^2 v^{-1} sec^{-1}$	Effective valence (Z^o)	Reference
Fe	trace	500	solid	A	2.6×10^{-8} (a)	-	-	122
C	trace	500	"	C	-	-	-	122
C	trace	500	"	C	-	-	-	170
B	-	-	"	C	-	-	-	170

(a) Units are in gm. atoms per sec. per amp. per ohm - cm.

Group III B Solvents

SOLVENT Al

Solute	Atomic % solute	Temp. (^{o}C.)	State	Solute accumulates at:	Transference no. (gm ion Faraday^{-1})	Electric mobility $cm^2 v^{-1} sec^{-1}$	Effective valence (Z^o)	Reference
Cu	-	-	liquid	C	-	-	-	71
Cu	50	1050	"	C	-	-	-	1,75
Cu	2,4,8 wt.%	100-200	solid (thin film)	A	-	-	-	223
Zn	0.84	560	solid	A	-	-	-11	63
Zn	5.0	577	"	A	-	-	-12	63
Ag	20 to 50 wt.%	900	liquid	A	-	-	-	1,40,75
Ag	trace	599	solid	A	-	-	-17	63
Au	9 wt.%	700	liquid	A	-	-	-	1,75
Si	dilute	-	"	A	-	-	-	185
Ge	dilute	-	"	A	-	-	-	185
H	-	-	"	C	-	-	-	11
Al (SelfET)	-	100 to 150	solid	A	(electron microscope thin foil)	-	-	135,136, 137,138
Al (SelfET)	-	150 to 450	"	A	(electron microscope thin foil)	-	-	192
Al (SelfET)	-	400 to 600	"	A	-	-	-	85
Al (SelfET)	-	500	"	A	-	-	-28	92

SOLVENT Al (Continued)

Solute	Atomic % solute	Temp. (°C.)	State	Solute accumulates at:	Transference no. (gm ion $Faraday^{-1}$)	Electric mobility $cm^2v^{-1}sec^{-1}$	Effective valence (Z^o)	Reference
Al (SelfET)	-	550	solid	A	-	-	-19	92
Al (SelfET)	-	360-605	"	A	-	-	-	212
Al (SelfET)	-	570	"	A	-	-	-16	92
Al (SelfET)	-	460 to 640	"	A	-	-	-12 to -30	115
Al (SelfET)	-	448 to 608	"	A	-	-	-	127
Al (SelfET)	-	90 to 100	"	A	-	-	-	173
Al (SelfET)	-	250 to 310	"	A	-	-	-	174

SOLVENT Ga

Solute	Atomic % solute	Temp. ($^oC.$)	State	Solute accumulates at:	Transference no. (gm ion $Faraday^{-1}$)	Electric mobility $cm^2v^{-1}sec^{-1}$	Effective valence (Z^o)	Reference
Hg	5 wt. %	30	liquid	A	-	($K = 1.9 \times 10^{+4}$)	-1.5	101
Sn	4 wt. %	30	"	A	-	($K = 1.7 \times 10^{+4}$)	-1.6	101
Sn	0.35 to 2.0 wt. %	100	"	A	-	2.54×10^{-3}	-	101
Bi	0.02 + 0.4 wt. %	100	"	A	-	(3 to 6) $\times 10^{-3}$	-	160
Si	dilute	-	"	A	-	-	-	185
Ge	dilute	-	"	A	-	-	-	185

SOLVENT In

Solute	Atomic % solute	Temp. ($^oC.$)	State	Solute accumulates at:	Transference no. (gm ion Faraday^{-1})	Electric mobility $cm^2v^{-1}sec^{-1}$	Effective valence (Z^o)	Reference
Ag	trace	-	liquid	A	-	-	-0.74	7,12
Ag	4.45 wt.%	168	"	A	-	-	-0.9	181
Ag	5.83 wt.%	176	"	A	-	-	-1.4	181
Ag	5.40 wt.%	178	"	A	-	-	-0.9	181
Sn	4.50 wt.%	166	"	A	-	$(1.61$ to $1.75) \times 10^{-3}$	-2.8 to -3.0	181
Pb	4.00 wt.%	193	"	A	-	1.74×10^{-3}	-2.3 to -2.4	181
Pb		350	"	A	-	$(1.95$ to $2.56) \times 10^{-3}$	-2.4 to -2.7	181
Bi	4.00 wt.%	166	"	A	-	3.43×10^{-3}	-6.5	181
Bi		170	"	A	-	3.60×10^{-3}	-	181
Bi		314	"	A	-	4.25×10^{-3}	-5.9	181
Tl	4.00 wt.%	172	"	-	-	-	<0.9	181
Tl	trace	350	"	A	-	-	-1.5	69,75
Co	trace	350	"	A	-	-	-12	69,75,76
Ni	trace	350	"	A	-	-	-4.7	69,75,76
Sb	50	684	"	A	-	0.82×10^{-4}	+1.34	166
Si	dilute	-	"	A	-	-	-	185
Ge	dilute	-	"	A	-	-	-	185
In (Self ET)		112	solid	A(a)	2.76×10^{-9}	-	-10.2	102
In (Self ET)		134	solid	A(a)	8.90×10^{-9}	-	-9.2	102

(a) Defects or impurities possibly to C.

SOLVENT Tl

Solute	Atomic % solute	Temp. (°C.)	State	Solute accumulates at:	Transference no. (gm ion $Faraday^{-1}$)	Electric mobility $cm^2 v^{-1} sec^{-1}$	Effective valence (Z^o)	Reference
Ag	trace	-	liquid	C	-	-	+0.64	7,12
Au	trace	-	"	A	-	-	-0.28	7,12
S	($Tl+Tl_2S$)	500 to 550	"	C(Tl)	-	-	Z^o_{Tl} = +0.26 to 0.37	221
Se	($Tl+Tl_2Se$)	450 to 550	"	C(Tl)	-	-	Z^o_{Tl} = +0.7 to +0.6	221

Group IV A Solvents

SOLVENT Ti

Solute	Atomic % solute	Temp. (^{o}C.)	State	Solute accumulates at:	Transference no. (gm ion Faraday^{-1})	Electric mobility $cm^2 v^{-1} sec^{-1}$	Effective valence (Z^o)	Reference
O	2	1130 to 1380	solid	A	-	(0.8 to 5) x 10^{-3}	-	22,35
C	dilute	950 to 1650	"	C	-	-	-	12,56
Ti (SelfET)	-	900 to 1060	"	A	-	-	-0.7[a] $Z^o/_f = -1.4$	111

(a) Assuming correlation factor, f = 0.5

SOLVENT Zr

Solute	Atomic % solute	Temp. ($^oC.$)	State	Solute accumulates at:	Transference no. (gm ion Faraday^{-1})	Electric mobility $cm^2v^{-1}sec^{-1}$	Effective valence (Z^o)	Reference
C	-	-	solid	C	-	-	-	169
O	2	1783	"	A	-	-	-	169,22,34
Zr (SelfET)	-	900 to 1200	"	A	-	-	-0.7[a] $Z^o/_f$ = -1.4	111
Zr (SelfET)	-	950 to 1740	"	C	-	-	+0.06[b] $Z^o/_f$ = 0.3 (±0.1)	211

(a) Assuming correlation factor, f = 0.5

(b) Assuming correlation factor, f = 0.2

Group IV B Solvents

SOLVENT Si

Solute	Atomic % solute	Temp. ($^oC.$)	State	Solute accumulates at:	Transference no. (gm ion Faraday^{-1})	Electric mobility $cm^2v^{-1}sec^{-1}$	Effective valence (Z^o)	Reference
B	trace	880 to 1380	solid	C	-	8.7×10^{-11} to 1.8×10^{-7}	-	85
Al	trace	1073 to 1240	"	C	-	1.01×10^{-8} to 1.1×10^{-6}	-	85
P	trace	850 to 925	"	C	-	2.53×10^{-11} to 1.16×10^{-10}	-	85
P	trace	925 to 1400	"	A	-	7.36×10^{-10} to 1.35×10^{-6}	(reversal)	85
Li	trace	360 to 860	"	C	-	$(0.258$ to $36.0) \times 10^{-6}$	-	28
Cu	trace	1000	"	C	-	-	+0.83	65
Cu	trace	1125	"	C	-	-	+1.23	65
Cu	trace	1200	"	C	-	-	+1.03	65
Fe	trace	1100	"	no effect	-	0	0	65
Au	trace	1100	"	no effect	-	0	0	65
Au	trace	1160	"	C	-	1×10^{-6}	-	67
Au	trace	1230	"	C	-	10×10^{-6}	-	67
Au	trace	1350	"	A	-	100×10^{-6}	(i.e.reversal)	67
Au	trace	900 to 1300	"	A	-	-	-	178
Zn	-	980 to 1270	"	C	-	-	+1.3 to +2.7 (mean = 1.97)	86
In	trace	900	"	C	-	-	+58	126
Ag	trace	840	"	C	-	-	+93	126

SOLVENT Ge

Solute	Atomic % solute	Temp. (oC.)	State	Solute accumulates at:	Transference no. (gm ion $Faraday^{-1}$)	Electric mobility $cm^2 v^{-1} sec^{-1}$	Effective valence (Z^o)	Reference
In	trace	300 to 550	solid	A	-	$(0.15$ to $3.6) \times 10^{-6}$	-	68
In	trace	800 to 900	"	C	-	$(9.8$ to $50) \times 10^{-6}$	(reversal)	68
Sb	trace	500 to 600	"	C	-	$(1.6$ to $4.2) \times 10^{-6}$	-	68
Sb	trace	800 to 900	"	A	-	$(12$ to $37) \times 10^{-6}$	(reversal)	68
Li	trace	150 to 600	"	C	-	$(0.085$ to $43.0) \times 10^{-6}$	-	28
Cu	trace	800 to 900	"	C	-	-	-	28
Cu	trace	700	"	C	-	-	-	65
Au	-	approx. 500	liquid zone	A	-	-	-	98, 99
Al	trace	1030	liquid	C	-	5.3×10^{-3}	+4.0	133
Ga	trace	1030	"	C	-	3.6×10^{-3}	+1.7	133
As	trace	1030	"	A	-	4.4×10^{-3}	-1.5	133
Ge (SelfET)	-	750	solid	C	-	-	+80	125

SOLVENT Sn

Solute	Atomic % solute	Temp. (°C.)	State	Solute accumulates at:	Transference no. (gm ion $Faraday^{-1}$)	Electric mobility $cm^2 v^{-1} sec^{-1}$	Effective valence (Z^o)	Reference
Co	trace	350	liquid	A	-	-	-6	69,75,76
Co	-	-	"	A	-	-	-	14
Sb	trace	455 to 465	"	A	-	-	+2.8 to +3.9	15,72
Zn	-	-	"	C	-	-	-	1
Ga	0.17	300	"	C	-	-	+0.6	8,9
Tl	trace	350	"	A	-	-	-1.3	69,75
Ni	trace	350	"	A	-		-2.8	69,75,76
Bi	2.85	350	"	A	-	-	-0.8	8,9,13
Bi	0.986	257	"	A	-	1.522×10^{-3}	-	26
Bi	0.986	378	"	A	-	1.741×10^{-3}	-	26
Bi	0.986	518	"	A	-	2.060×10^{-3}	approx. -1.95	26
Bi	4.98	250	"	A	-	1.488×10^{-3}	-	26
Bi	4.98	351	"	A	-	1.675×10^{-3}	approx. -2.225	26
Bi	4.98	450	"	A	-	1.790×10^{-3}	approx. -1.92	26
Bi	4.98	518	"	A	-	1.863×10^{-3}	approx. -1.8	26
Bi	4.98	599	"	A	-	1.923×10^{-3}	approx. -1.7	26
Bi	9.96	516	"	A	-	1.801×10^{-3}	approx. -1.725	26
Bi	20.12	220	"	A	-	1.283×10^{-3}	-	26
Bi	20.12	338	"	A	-	1.506×10^{-3}	approx. -2.06	26
Bi	20.12	518	"	A	-	1.687×10^{-3}	approx. -1.46	26

SOLVENT Sn (Continued)

Solute	Atomic % solute	Temp. (°C.)	State	Solute accumulates at:	Transference no. (gm ion $Faraday^{-1}$)	Electric mobility $cm^2 v^{-1} sec^{-1}$	Effective valence (Z^o)	Reference
Bi	38.57	177	liquid	A	-	0.993×10^{-3}	-	26
Bi	38.57	297	"	A	-	1.223×10^{-3}	approx. -1.475	26
Bi	38.57	517	"	A	-	1.475×10^{-3}	approx. -1.1	26
Bi	1.0	430	"	A? (not stated)	-	$(1.52 \text{ to } 1.79) \times 10^{-3}$ dep. on current density	-	100
Bi	4.8	350	"	A	-	-	-0.8	168
Bi	43.0	222	"	A	-	-	-	106
Bi	0 to 80	300	"	A	-	-	-2.56 to -0.40	219
Au	trace	-	"	A	-	-	-2.2	7,12
Ag	16 wt.%	-	"	C	-	-	-	1
Ag	trace	-	"	A	-	-	-1.07	1,7,12
Ag	0.0002 to 0.2	380 to 390	"	A	-	$(3.3 \text{ to } 3.9) \times 10^{-4}$	-0.39 to -0.47	38
Ag	0.0002 to 0.2	630	"	A	-	$(5.8 \text{ to } 7.1) \times 10^{-4}$	-0.48 to -0.57	38
Al	0.002 to 2.0	380 to 390	"	C	-	$(5.1 \text{ to } 6.6) \times 10^{-4}$	+0.55 to +0.7	38
Al	0.002 to 2.0	630	"	C	-	$(7.9 \text{ to } 9.9) \times 10^{-4}$	+0.58 to +0.73	38
Al	42	800 to 1600	"	C	-	-	-	1,75
Mn	0.002 to 2.0	380 to 390	"	A	-	$(5.8 \text{ to } 7.2) \times 10^{-4}$	-0.66 to -0.81	38
Mn	0.002 to 2.0	630	"	A	-	$(15.2 \text{ to } 18.3) \times 10^{-4}$	-1.16 to -1.4	38
Si	dilute	-	"	A	-	-	-	185
Ge	0.002 to 2.0	380 to 390	"	A	-	$(3.6 \text{ to } 4.1) \times 10^{-4}$	-0.45 to -0.51	38

SOLVENT Sn (Continued)

Solute	Atomic % solute	Temp. (°C.)	State	Solute accumulates at:	Transference no. (gm ion $Faraday^{-1}$)	Electric mobility $cm^2 v^{-1} sec^{-1}$	Effective valence (Z^o)	Reference
Ge	0.002 to 2.0	630	liquid	A	-	$(8.9 \text{ to } 11.1) \times 10^{-4}$	-0.73 to -0.98	38
Ge	dilute	-	"	A	-	-	-	185
Cd	-	300	"	C	-	-	-	1
Ca	0.1	300	"	C	-	-	+0.75	168
Cd	20-30 wt.%	330	"	C	-	-	-	1,71,75
Cu	3 wt.%	370 to 420	"	C	-	-	-	71
Cu	-	1000	"	C	-	-	-	1
Pb	12.5,26.1	234,211 to 240	"	A	-	-	-	166
Pb	37 wt.%	350	"	A	-	-	-	1,75
Pb	0 to 80	350	"	A	-	-	-1.6 to -0.13	219
Sn (SelfET)	-	190	solid	A	-	-	-80	12,92

SOLVENT Pb

Solute	Atomic % solute	Temp. (^{o}C.)	State	Solute accumulates at:	Transference no. (gm ion Faraday^{-1})	Electric mobility $cm^2 v^{-1} sec^{-1}$	Effective valence (Z^o)	Reference
Cd	-	300	liquid	C	-	-	-	1
Sb	1.38 to 1.55	350 to 695	"	A	-	-	-	167
Sb	13.5 wt.%	470	"	A	-	-	-	1,71,75
Ag	2 wt.%	360 to 390	"	C	-	-	-	71
Ag	approx. 0.02	360	"	C	-	-	+0.33	8,9,14
Ag	trace	-	"	C	-	-	+0.48	7,12
Ag	37 to 70	1000	"	C	-	-	-	213
Ag	0.04	200, 290	solid	A	-	-	-	1
Au	50 wt.%	450	liquid	C	-	-	-	1
Au	trace	-	"	A	-	-	-0.1	7,12
Au	0.04	200	solid	A	1×10^{-10}	2.5×10^{-6}	-	1,22,32
Au	0.04	265	"	A	5.1×10^{-10}	-	-	1,33
		280	"	A	9.1×10^{-10}	-	-	1,33
		290	"	A	1.54×10^{-9}	1.5×10^{-5}	-	1,22,33
Zn	0.13	360	liquid	C	-	-	+0.5 to +0.9	8,9
Bi	-	-	"	A	-	-	-	1
Bi	0 to 80	500	"	A	-	-	-0.84 to -0.26	219
Se	0.22	360	"	A	-	-	-2.0	8,9
Sn	-	-	"	C	-	-	-	11,13
In	33.5	312	"	C	-	-	-	166

SOLVENT Pb (Continued)

Solute	Atomic % solute	Temp. ($^oC.$)	State	Solute accumulates at:	Transference no. (gm ion $Faraday^{-1}$)	Electric mobility $cm^2v^{-1}sec^{-1}$	Effective valence (Z^o)	Reference
Tl	trace	350	liquid	C	-	-	+0.4	69,75
Co	trace	350	"	A	-	-	-11	69,75,76
Pb (SelfET)	-	250	solid	A	-	-	-47	92
Pb (SelfET)	-	310	"	A	-	-	-45	92

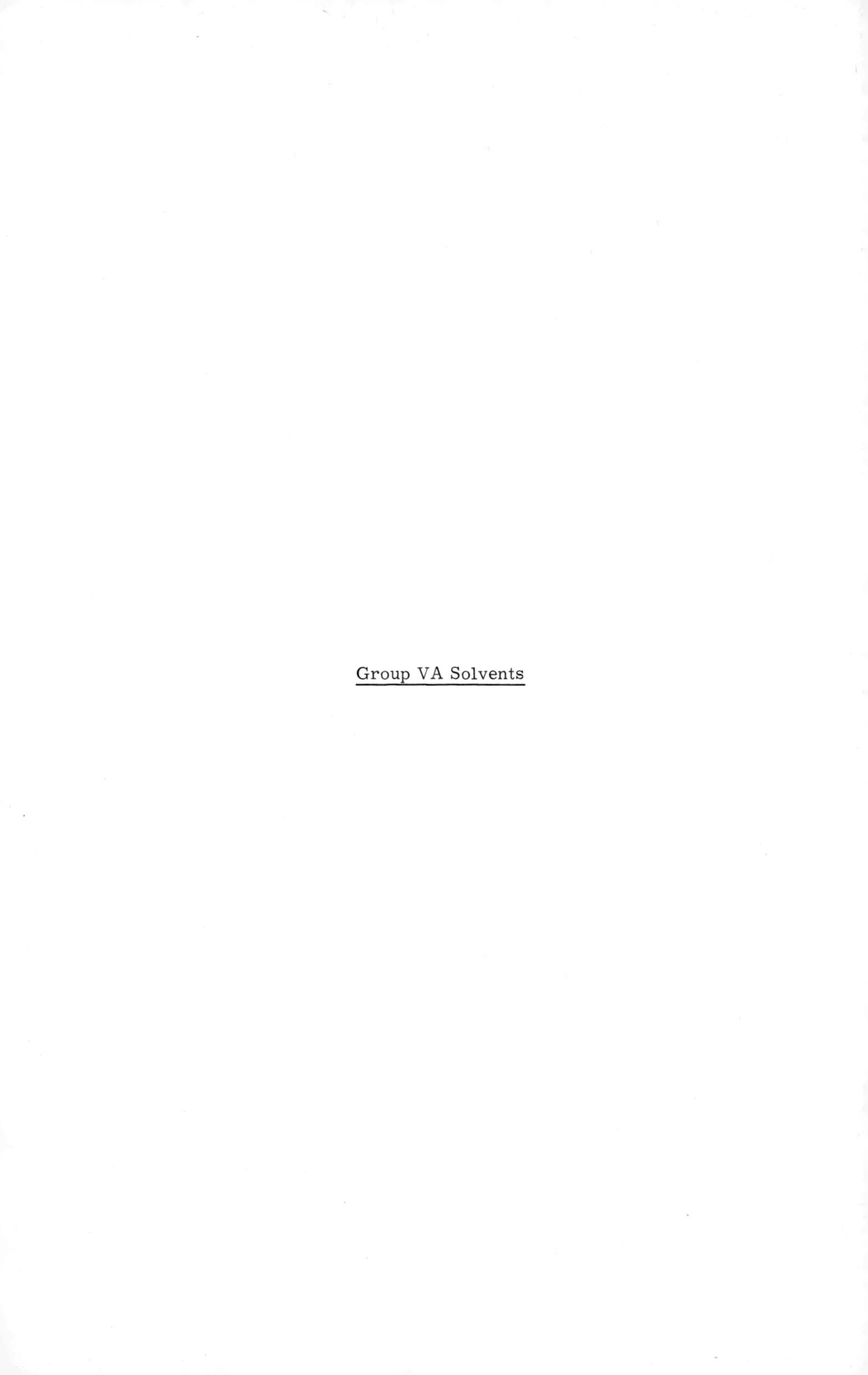

Group VA Solvents

SOLVENT V

Solute	Atomic % solute	Temp. ($^oC.$)	State	Solute accumulates at:	Transference no. (gm ion $Faraday^{-1}$)	Electric mobility $cm^2 v^{-1} sec^{-1}$	Effective valence (Z^o)	Reference
C	trace	1650	solid	C	-	10.7×10^{-5}	+1.9	128
C	trace	1755	"	C	-	16.9×10^{-5}	+2.5	128
C	trace	1825	"	C	-	22.0×10^{-5}	+2.1	128
N	trace	1650	"	C	-	$(3.7 \text{ and } 3.9) \times 10^{-5}$	+1.6 and +1.9	128
N	(25 ppm)	1755	"	C	-	$(6.4 \text{ and } 6.5) \times 10^{-5}$	+1.6 and +1.7	128
N		1825	"	C	-	$(8.4 \text{ and } 8.0) \times 10^{-5}$	+1.7 and +1.8	128
O	trace	1650	"	C	-	7.8×10^{-5}	+1.3	128
O	(245 ppm)	1735	"	C	-	8.2×10^{-5}	+1.0	128
O		1825	"	C	-	17.5×10^{-5}	+1.3	128

SOLVENT Ta

Solute	Atomic % solute	Temp. ($^oC.$)	State	Solute accumulates at:	Transference no. (gm ion $Faraday^{-1}$)	Electric mobility $cm^2 v^{-1} sec^{-1}$	Effective valence (Z^o)	Reference
C	dilute	600 to 2600	solid	C	-	-	-	12,56

Group VB Solvents

SOLVENT Sb

Solute	Atomic % solute	Temp. (°C.)	State	Solute accumulates at:	Transference no. (gm ion Faraday^{-1})	Electric mobility $cm^2 v^{-1} sec^{-1}$	Effective valence (Z^o)	Reference
Ag	10 wt. %	-	liquid	C	-	-	-	1
Au	50 wt. %	400	"	C	-	-	-	1
Zn	20 to 60	500	"	C	-	-	-	1
Zn	39 to 47 wt. %	620	"	C	-	-	-	1, 75
Pb	1.26 to 1.50	500	"	C	-	-	-	167

SOLVENT Bi

Solute	Atomic % solute	Temp. ($^{o}C.$)	State	Solute accumulates at:	Transference no. (gm ion $Faraday^{-1}$)	Electric mobility $cm^2v^{-1}sec^{-1}$	Effective valence (Z^o)	Reference
Ag	30 wt. %	-	liquid	C	-	-	-	1,75
Ag	approx. 0.01	300	"	C	-	-	+0.1 to 0.2	8,9
Ag	trace	-	"	C	-	-	+0.2	7,12
Ag	0.94, 0.57 wt. %	305 to 605	"	C	-	$(0.75 \text{ to } 2.4) \times 10^{-4}$	+0.1 to 0.275	39
Ag	trace	475 to 515	"	C	-	-	-	1,15
Ag	dilute	500	"	C	-	2.2×10^{-4}	-	39,94
Sb	25 to 88 wt. %	300	solid	C	-	-	-	1,75
U	0.113 wt. %	478	liquid	A	-	-	-	25
Cu	0.49 wt. %	495	"	C	-	-	-	25,94
Mg	0.508 wt. %	345	"	C	-	-	-	25
Zr	0.0567 wt. %	510	"	A	-	-	-	25
Pd	0.188 wt. %	453	"	A	-	-	-	25
Ni	0.486 wt. %	500	"	A	-	-	-	25
Tl	trace	350	"	C	-	-	+0.2	69,75
Co	trace	350	"	A	-	-	-1.2	69,75,76
Ni	trace	350	"	A	-	-	-0.7	69,75,76
Au	50 wt. %	680	"	C	-	-	-	1,75
Au	trace	-	"	A	-	-	-0.18	7,12
Cd	25 to 75 wt. %	-	"	C	-	-	-	1,12,75

SOLVENT Bi (Continued)

Solute	Atomic % solute	Temp. (^oC.)	State	Solute accumulates at:	Transference no. (gm ion Faraday^{-1})	Electric mobility $cm^2v^{-1}sec^{-1}$	Effective valence (Z^o)	Reference
Cd	2.53	300	liquid	C	-	-	+1.36	8, 9, 12, 13, 71
Cd	0.32 wt. %	500	"	C	-	8.6×10^{-4}	-	94, 97
Cd	0.38 wt. %	450	"	C	-	9.5×10^{-4}	-	94, 97
Cd	0.42 wt. %	305 to 515	"	C	-	$(6.5 \text{ to } 8.5) \times 10^{-4}$	-	94, 97
Cd	0.82 wt. %	495 and 535	"	C	-	11.1×10^{-4}	-	94, 97
Cd	0.88 wt. %	342	"	C	-	7.3×10^{-4}	-	94, 97
Cd	0.90 wt. %	450	"	C	-	9.5×10^{-4}	-	94, 97
Cd	0.94 wt. %	320	"	C	-	6.0×10^{-4}	-	94, 97
Cd	0.95 wt. %	388	"	C	-	6.9×10^{-4}	-	94, 97
Cd	1.05 wt. %	310	"	C	-	10.6×10^{-4}	-	94, 97
Se	0.085 wt. %	300	"	A	-	-	-0.9	8, 9
Cr	-	-	"	A	-	-	-	12
Fe	-	-	"	C	-	-	-	12
Te	-	-	"	A	-	-	-	13
Pb	43.6	225 to 300	"	C	-	-	-	1, 71, 75
Pb	40 wt. %	240 to 400	"	C	-	-	-	1
Sn	5.16	518	"	C	-	0.78×10^{-3}	approx. +1.05	26
Sn	21.2	248	"	C	-	0.691×10^{-3}	-	26

SOLVENT Bi (Continued)

Solute	Atomic % solute	Temp. (°C.)	State	Solute accumulates at:	Transference no. (gm ion Faraday^{-1})	Electric mobility $cm^2v^{-1}sec^{-1}$	Effective valence (Z^o)	Reference
Sn	21.2	368	liquid	C	-	0.823×10^{-3}	approx. +1.35	26
Sn	21.2	520	"	C	-	0.943×10^{-3}	approx. +0.95	26
Sn	42.41	197	"	C	-	0.858×10^{-3}	-	26
Sn	42.41	318	"	C	-	1.024×10^{-3}	approx. +1.325	26
Sn	42.41	520	"	C	-	1.204×10^{-3}	approx. +0.825	26
Sn	-	200, 300	"	C	-	-	-	1
Sn	-	-	"	C	-	-	-	71
Sn	0.88 wt. %	510	"	C	-	7.7×10^{-4}		94, 95
Sn	0.88 wt. %	430	"	C	-	6.43×10^{-4}	-	95
Sn	0.88 wt. %	310	"	C	-	5.61×10^{-4}	-	95
Sn	0.89 wt. %	600	"	C	-	7.73×10^{-4}	-	95
Sn	0.86 wt. %	355	"	C	-	6.03×10^{-4}	-	95
In	0.78 wt. %	358	"	C	-	7.4×10^{-4}	-	94, 97
In	0.85 wt. %	387	"	C	-	8.4×10^{-4}	-	94, 97
In	0.86 wt. %	320	"	C	-	8.1×10^{-4}	-	94, 97
In	0.87 wt. %	355 and 410	"	C	-	$(7.7 \text{ and } 8.8) \times 10^{-4}$	-	94, 97
In	0.88 wt. %	480	"	C	-	8.5×10^{-4}	-	94, 97
In	0.91 wt. %	405	"	C	-	7.5×10^{-4}	-	94, 97
In	0.94 wt. %	595	"	C	-	9.3×10^{-4}	-	94, 97

SOLVENT Bi (Continued)

Solute	Atomic % solute	Temp. ($^oC.$)	State	Solute accumulates at:	Transference no. (gm ion $Faraday^{-1}$)	Electric mobility $cm^2v^{-1}sec^{-1}$	Effective valence (Z^o)	Reference
In	0.95 wt.%	497	liquid	C	-	8.9×10^{-4}	-	94,97
In	1.05 wt.%	310	"	C	-	7.0×10^{-4}	-	94,97
Zn	15	700	"	C	-	-	0.417	189
Zn	30	700	"	C	-	-	0.704	189
Zn	50	700	"	C	-	-	1.84	189
Zn	70	700	"	C	-	-	6.12	189
Zn	85	700	"	C	-	-	9.53	189
Sb	1.05 wt.%	510	"	C	-	1.23×10^{-4}	-	94,95
Sb	0.93 wt.%	562	"	C	-	1.41×10^{-4}	-	95
Sb	0.93 wt.%	412	"	C	-	1.06×10^{-4}	-	95
Sb	0.93 wt.%	310	"	C	-	0.98×10^{-4}	-	95
Sb	0.93 wt.%	605	"	C	-	1.15×10^{-4}	-	95
Sb	0.93 wt.%	493	"	C	-	1.23×10^{-4}	-	94,95

Group VI A Solvents

SOLVENT Cr

Solute	Atomic % solute	Temp. (°C.)	State	Solute accumulates at:	Transference no. (gm ion $Faraday^{-1}$)	Electric mobility $cm^2 v^{-1} sec^{-1}$	Effective valence (Z^o)	Reference
Fe	dilute	900 to 1200	solid	A	-	-	-	12,50,53

SOLVENT Mo

Solute	Atomic % solute	Temp. (°C.)	State	Solute accumulates at:	Transference no. (gm ion $Faraday^{-1}$)	Electric mobility $cm^2 v^{-1} sec^{-1}$	Effective valence (Z^o)	Reference
W	25 and 75	1500 to 2200	solid	C(Mo to A)	-	-	-	57,58
Fe	dilute	950 to 1100	"	C	-	-	-	12,50
Ni	90.76 wt.%	1150	"	A	-	-	-25.7	12,46
Ni	90.76 wt.%	1200	"	A	-	-	-20.9	12,46
Ni	90.76 wt.%	1250	"	A	-	-	-18.0	12,46
Ni	90.76 wt.%	1300	"	A	-	-	-15.0	12,46
Cr	9.92 wt.%	1200 to 1350	"	C	-	-	-	48
Mo (SelfET)	-	1650 to 1900	"	C	-	-	-32	183

SOLVENT W

Solute	Atomic % solute	Temp. (°C.)	State	Solute accumulates at:	Transference no. (gm ion $Faraday^{-1}$)	Electric mobility $cm^2v^{-1}sec^{-1}$	Effective valence (Z^o)	Reference
Th	surf. trace	450	solid	C(W to C)	-	-	-	1,24
Mo	dilute	-	liquid	A	-	-	-	20
Mo	25 and 75	1900 to 2500	solid	A(W to C)	-	-	-	57,58
C	dilute	1800 to 2800	"	C	-	-	-	12,56
Fe	dilute	900 to 1150	"	A	-	-	-	12,50,51
W (SelfET)	-	1950	"	C	-	approx.	+25.2	107
W (SelfET)	-	1955	"	C	-	(15 to 25) $x 10^{-12}$	+28.1	107
W (SelfET)	-	2020	"	C	-	$cm^3A^{-1}sec^{-1}$	+17.3	107
W (SelfET)	-	2080	"	C	-		+10.7	107
W (SelfET)	-	2095	"	C	-		+12.4	107
W (SelfET)	-	1697 to 1762	"	C	-	-	-	191

Group VI B Solvents

SOLVENT Se

Solute	Atomic % solute	Temp. ($^{o}C.$)	State	Solute accumulates at:	Transference no. (gm ion $Faraday^{-1}$)	Electric mobility $cm^2 v^{-1} sec^{-1}$	Effective valence (Z^o)	Reference
Tl	trace	100	solid	C	-	-	+2.85	66
Tl	trace	150	"	C	-	-	+2.81	66
Tl	trace	175	"	C	-	-	+3.59	66
Tl	trace	200	"	C	-	-	+3.31	66
Tl	trace	215	"	C	-	-	+3.34	66

SOLVENT Te

Solute	Atomic % solute	Temp. ($^oC.$)	State	Solute acc mulates at:	Transference no. (gm ion Faraday^{-1})	Electric mobility $cm^2v^{-1}sec^{-1}$	Effective valence (Z^o)	Reference
Tl	trace	360	solid	C	-	1.7×10^{-10}	+3.5(a)	87
Tl	trace	380	"	C	-	4.4×10^{-10}	+4.3(a)	87
Tl	trace	380	"	C	-	0.32×10^{-10}	+7.5(b)	87
Tl	trace	380	"	C	-	0.55×10^{-10}	+3.8(c)	87
Tl	trace	400	"	C	-	8.6×10^{-10}	+5.3(a)	87
Tl	trace	400	"	C	-	3.0×10^{-10}	+10.7(b)	87
Tl	trace	400	"	C	-	7.3×10^{-10}	+5.5(c)	87
Tl	trace	430	"	C	-	59.8×10^{-10}	+6.1(a)	87
Tl	trace	430	"	C	-	43.6×10^{-10}	+18.4(b)	87
Tl	trace	430	"	C	-	90.0×10^{-10}	+6.2(c)	87
Cd	0.1 to 1.5	470 to 560	liquid	C	-	-	$+1.10 \pm 0.07$	219

(a) Polycrystalline specimen.

(b) Single crystal specimen parallel to C-axis.

(c) Single crystal specimen perpendicular to C-axis.

Group VIII Solvents

SOLVENT Fe

Solute	Atomic % solute	Temp. (^{o}C.)	State	Solute accumulates at:	Transference no. (gm ion Faraday^{-1})	Electric mobility $cm^2 v^{-1} sec^{-1}$	Effective valence (Z^{o})	Reference
C	4.5	1000	solid	C	1.6×10^{-6}	2.2×10^{-5}	-	1,22,30
C	2	1175 to 1277	"	C	$(2.4 \text{ to } 5.7) \times 10^{-6}$	$(2.0 \text{ to } 5.8) \times 10^{-5}$	+0.9 to 3.9	22,31
C	4	1124 to 1226	"	C	$(12.7 \text{ to } 14.2) \times 10^{-6}$	$(4.9 \text{ to } 5.6) \times 10^{-5}$	+2.7 to 4.9	22,31
C	2.6 wt.%	1300	liquid	C	-	-	-	1,71,75
C	4.5	1000	solid	C	-	-	-	1
C	(α-Fe)	600	"	C	-	-	+49.3	41,44,84
C	(γ-Fe, 1%C)	950	"	C(Fe to A)	for C 3.06×10^{-6}	-	+13.4	41,44,49
C	(γ-Fe, 1%C)	1000	"	C(Fe to A)	for C 4.21×10^{-6}	-	+11.0	41,44,49
C	(γ-Fe, 1%C)	1050	"	C(Fe to A)	for C 4.85×10^{-6}	-	+7.7	41,44,49
C	(γ-Fe, 1%C)	1100	"	C(Fe to A)	for C 10.35×10^{-6}	-	+10.6	41,44,49
C	(γ-Fe, 1%C)	1150	"	C(Fe to A)	for C 12.04×10^{-6}	-	+8.1	41,44,49
C	(γ-Fe, 1%C)	800 to 1400	"	C	-	-	-	12,50
C	1.6	1600	liquid	C	-	-	+11	77
C	(γ-Fe, 0.35 %C)	945	solid	C(Fe to A)	for Fe 6.4×10^{-6}	-	-	42
C	(γ-Fe, 0.35 %C)	1020	"	C(Fe to A)	for Fe 15×10^{-6}	-	-	42
C	(γ-Fe, 0.7 %C)	945	"	C(Fe to A)	for Fe 5.3×10^{-6}	-	-	42
C	(γ-Fe, 0.7 %C)	1020	"	C(Fe to A)	for Fe 6.9×10^{-6}	-	-	42
C	0.03 wt.% (α-Fe)	752	"	C	-	8×10^{-5}	+4.3	141

SOLVENT Fe (Continued)

Solute	Atomic % solute	Temp. ($^oC.$)	State	Solute accumulates at:	Transference no. (gm ion Faraday^{-1})	Electric mobility $cm^2v^{-1}sec^{-1}$	Effective valence (Z^o)	Reference
C	0.4 wt.%	842	solid	C	-	0.58×10^{-6}	+13.7	142
C	(γ-Fe)	902	"	C	-	0.92×10^{-6}	+9.3	142
C		925	"	C	-	2.05×10^{-6}	+8.85	142
C		950	"	C	-	1.1105×10^{-6}	+8.55	142
C	0.4 wt.% (+4.0% Mn)	950	"	C	-	1.61×10^{-6}	+9.15(a)	142
C	0.4 wt.% (+2.0% Si)	950	"	C	-	0.845×10^{-6}	+4.75(b)	142
C	0.11 to 0.5 wt.%(γ-Fe)	927 to 1027	"	C	-	-	+3.99	176
C	1.0 wt.%	1000	"	C	-	-	+11	177
C	0.5 wt.%	950	"	C	-	-	+10.8	180
C	0.5 wt.%	1400	"	C	-	-	+8.7	180
C	3.65 wt.% (1.72 Mn)	-	liquid	C	-	-	-	10
S	1.88 wt.% (0.07 P)	-	"	C	-	-	-	10
Al	25,35,40 at.%	1000 to 1340	solid	C(Fe to A)	-	-	-	12,61
Al	2.5 wt.%	1300	"	(Fe to C)	for Fe 4.6×10^{-6}	-	-	12,43
Al	8 wt.%	1300	"	(Fe to C)	for Fe 3.4×10^{-6}	-	-	12,43
W	0.49 wt.%	900 to 1150	"	C	for W (0 to 3.2) $\times 10^{-8}$	-	-	12,41,49, 50,51,52

SOLVENT Fe (Continued)

Solute	Atomic % solute	Temp. ($^oC.$)	State	Solute accumulates at:	Transference no. (gm ion $Faraday^{-1}$)	Electric mobility $cm^2v^{-1}sec^{-1}$	Effective valence (Z^o)	Reference
Si	dilute	-	liquid	C	-	-	-	185
Ge	dilute	-	"	C	-	-	-	185
H	dilute	441 to 550	solid	C	-	-	+0.22 to 0.26	118
Deuterium	"	463 and 556	"	C	-	-	+0.38 to +0.39	118
N	-	1000	"	A	-	3×10^{-5}	-	1,82
N	0.1 wt.% (α-Fe)	736	"	C	-	4.95×10^{-5}	+5.7	141
N	0.4 wt.% (γ-Fe)	922	"	A	-	5.4×10^{-6}	-14.1	142
N		962	"	A	-	7.05×10^{-6}	-11.55	142
N		1000	"	A	-	8.7×10^{-6}	-8.35	142
Cr	7.4 wt.%	900 to 1200	"	C	$(1.04 \text{ to } 2.26) \times 10^{-6}$	-	-	12,41,49, 50,53,168
Mo	2 wt.%	950 to 1100	"	A	$(1.01 \text{ to } 1.87) \times 10^{-6}$	-	-	12,41,49, 50,54
Mo	4.5 wt.%	950 to 1050	"	A	$(6.48 \text{ to } 12.32) \times 10^{-7}$	-	-	12,41,49, 50,54
Ni	-	-	"	no effect	0	0	0	1,82
Ni	8	1600	liquid	A	-	-	-4	75,77
B	-	1040	solid	C	-	1×10^{-5}	-	1,82
Mn	0.5	1600	liquid	C	-	-	+5	75,77

SOLVENT Fe (Continued)

Solute	Atomic % solute	Temp. (°C.)	State	Solute accumulates at:	Transference no. (gm ion $Faraday^{-1}$)	Electric mobility $cm^2 v^{-1} sec^{-1}$	Effective valence (Z^o)	Reference
Fe (SelfET)	-	1050 to 1300	solid	C	-	-	-	36,85
Fe (SelfET)	-	1190	"	C	1.8×10^{-8}	-	+43	91
Fe (SelfET)	-	1300	"	C	$(4.2 \text{ to } 5.8) \times 10^{-8}$	-	+12	91
Fe (SelfET)	-	1325	"	C	-	-	+9	91
Fe (SelfET)	-	906 to 1385	"	C	-	-	+1 to +3	103,113

(a) The presence of Mn increases the effective valence of C in Fe.

(b) The presence of Si decreases the effective valence of C in Fe.

SOLVENT Co

Solute	Atomic % solute	Temp. (^{o}C.)	State	Solute accumulates at:	Transference no. (gm ion Faraday^{-1})	Electric mobility cm^{2}v^{-1}sec^{-1}	Effective valence (Z^{o})	Reference
Si	dilute	-	liquid	C	-	-	-	185
Ge	dilute	-	"	C	-	-	-	185
W	0.82 wt.%	1100 to 1350	solid	C	-	-	>+10	55
C	-	600 to 1400	"	C	-	-	-	12,50,44
C	0.05 to 0.12 wt.%	800 to 1300 approx.	"	C	-	-	+6.5 to + 10 approx.	104
Co (SelfET)	-	1250 to 1360	"	C	-	-	+1.0 to + 2.2	114

SOLVENT Ni

Solute	Atomic % solute	Temp. (^oC.)	State	Solute accumulates at:	Transference no. (gm ion Faraday^{-1})	Electric mobility $cm^2v^{-1}sec^{-1}$	Effective valence (Z^o)	Reference
H	dilute	447 to 625	solid	C	-	-	+0.52 to +0.61	118
Deuterium	"	482 to 645	"	C	-	-	+0.67 to 0.76	118
Sb	28.3	650	"	only Ni to A	$0.68 \times 10^{+10}$(a)	-	-3.6	139
Sb	28.3	715	"	"	$1.2 \times 10^{+10}$(a)	-	-3.6	139
Sb	28.3	800	"	"	$2.3 \times 10^{+10}$(a)	-	-3.8	139
Sb	28.3	885	"	"	$3.2 \times 10^{+10}$(a)	-	-4.5	139
Sb	28.3	950	"	"	$5.5 \times 10^{+10}$(a)	-	-4.1	139
Sb	28.3	1030	"	"	$7.8 \times 10^{+10}$(a)	-	-4.1	139
Si	dilute	-	liquid	C	-	-	-	185
Ge	dilute	-	"	C	-	-	-	185
C	-	600 to 1400	solid	C	-	-	-	12,50
Cr	4.36 wt.%	950	"	C	-	-	+57.6	45
Cr	4.36 wt.%	1000	"	C	-	-	+42.5	45
Cr	4.36 wt.%	1050	"	C	-	-	+34.7	45
Cr	4.36 wt.%	1100	"	C	-	-	+27.6	45
Cr	4.76 to 31.01	1250	"	C(Ni to A)	-	-	-	57
W	dilute	850 to 1100	"	C	-	-	-	12,47
Fe	2	1300	liquid	C	-	-	+15	77
Mn	10	1500	"	C	-	-	+56 to + 76	77

SOLVENT Ni (Continued)

Solute	Atomic % solute	Temp. (°C.)	State	Solute accumulates at:	Transference no. (gm ion $Faraday^{-1}$)	Electric mobility $cm^2v^{-1}sec^{-1}$	Effective valence (Z^o)	Reference
Mo	8	1000to1400	solid	C(Ni to A)	-	-	Z^o_{Mo} +1.71 to +1.31	188,218
							Z^o_{Ni} -1.12 to -0.79	
Mo	16	950 to 1350	"	C(Ni to A)	-	-	Z^o_{Mo} +20.4 to +16.4	196,218
							Z^o_{Ni} -4.02 to -3.02	
Mo	18	950 to 1350	"	C(Ni to A)	-	-	Z^o_{Mo} +13.6 to +11.4	218
							Z^o_{Ni} -1.61 to -1.16	
Mo	20	950 to 1300	"	C(Ni to A)	-	-	Z^o_{Mo} +33.5 to +38.8	187,218
							Z^o_{Ni} -9.9 to -8.3	
Mo	23	1100 to 1300	"	C(Ni to A)	-	-	Z^o_{Mo} +14.0 to +12.5	197,218
							Z^o_{Ni} -4.46 to -4.01	
Ni (SelfET)	-	1000 to 1200	"	A	-	-	-	36
Ni (SelfET)	-	1190	"	A	(5 and 7) x 10^{-9}	-	-5	91
Ni (SelfET)	-	1255	"	A	6 x 10^{-9}	-	-2.1	91
Ni (SelfET)	-	1300	"	A	10 x 10^{-9}	-	-1.8	91

SOLVENT Ni (Continued)

Solute	Atomic % solute	Temp. ($^oC.$)	State	Solute accumulates at:	Transference no. (gm ion Faraday^{-1})	Electric mobility $cm^2v^{-1}sec^{-1}$	Effective valence (Z^o)	Reference
Ni (SelfET)	-	1385	solid	A	23×10^{-9}	-	-1.5	91
Ni (SelfET)	-	1000 to 1400	"	A		-	-3.2 to -3.5	121

(a) Units are mol. amp^{-1}sec^{-1}.

SOLVENT Pd

Solute	Atomic % solute	Temp. (°C.)	State	Solute accumulates at:	Transference no. (gm ion Faraday^{-1})	Electric mobility $cm^2v^{-1}sec^{-1}$	Effective valence (Z^o)	Reference
H	37	10 to 72	solid	C	-	0.4 - 2.7	-	1,78,80
H			"	C	-	-	-	195
H	2.8	182	"	C	0.79×10^{-6}	1.5×10^{-4}	-	19,22
H	2.0	240	"	C	1.19×10^{-6}	2.8×10^{-4}	-	19,22
Si	dilute	-	liquid	C	-	-	-	185
Ge	dilute	-	"	C	-	-	-	185

SOLVENT Pt

Solute	Atomic % solute	Temp. (°C.)	State	Solute accumulates at:	Transference no. (gm ion $Faraday^{-1}$)	Electric mobility $cm^2 v^{-1} sec^{-1}$	Effective valence (Z^o)	Reference
Si	dilute	-	liquid	C	-	-	-	185
Ge	dilute	-	"	C	-	-	-	185
Pt (SelfET)	-	1200 to1680	solid	C	-	-	+0.28 at 1590°C +0.32 at 1680°C	85,112
Pt (SelfET)	(Displacement of grain boundaries towards C)				-	-	-	120

Section B

Electrotransport in Ternary Alloys

COMPONENTS	RELATIVELY MIGRATING SOLUTE(S)	WEIGHT % SOLUTE(S)	TEMP. (°C)	STATE	SOLUTE(S) ACCUMULATE AT:	REFERENCE
Hg	-	-	240	liquid	-	1,200
Na	Na	40			C	
Sn	Sn	1 to 5			C	
Cu	-	-	1050	liquid	-	1,201
Be	Be	10			C	
Fe	(Fe)	-			(C)	
Pb	-	-	1000	liquid	-	1,202
Cu	Cu	-			C	
Zn	Zn	-			C	
Sn	-	7 to 11	1000	liquid	-	1,202,203
Cu	Cu	30			C	
Ag	Ag	-			C	
Sn	-	-	1000	liquid	-	1,200
Cu	Cu	30			C	
Bi	-	2 to 20			-	
Sn	-	-	1000	liquid	-	1,204
Cu	Cu	-			C	
Pb	-	1			-	

COMPONENTS	RELATIVELY MIGRATING SOLUTE(S)	WEIGHT % SOLUTE(S)	TEMP. (°C)	STATE	SOLUTE(S) ACCUMULATE AT:	REFERENCE
Sn	-	-	660	liquid	-	1,205
Al	-	-			-	
Fe	Fe	-			C	
Bi	-	-	475	liquid	-	25
U	U	0.150			A	
Cu	Cu	0.480			C	
Bi	-	-	344	liquid	-	25
Mg	Mg	0.262			C	
Cu	Cu	0.372			C	
Bi	-	-	499	liquid	-	25
Ni	Ni	0.497			A	
Cu	Cu	0.493			C	
Bi	-	-	350	liquid	-	25
Ni	Ni	0.286			A	
Mg	Mg	0.269			C	

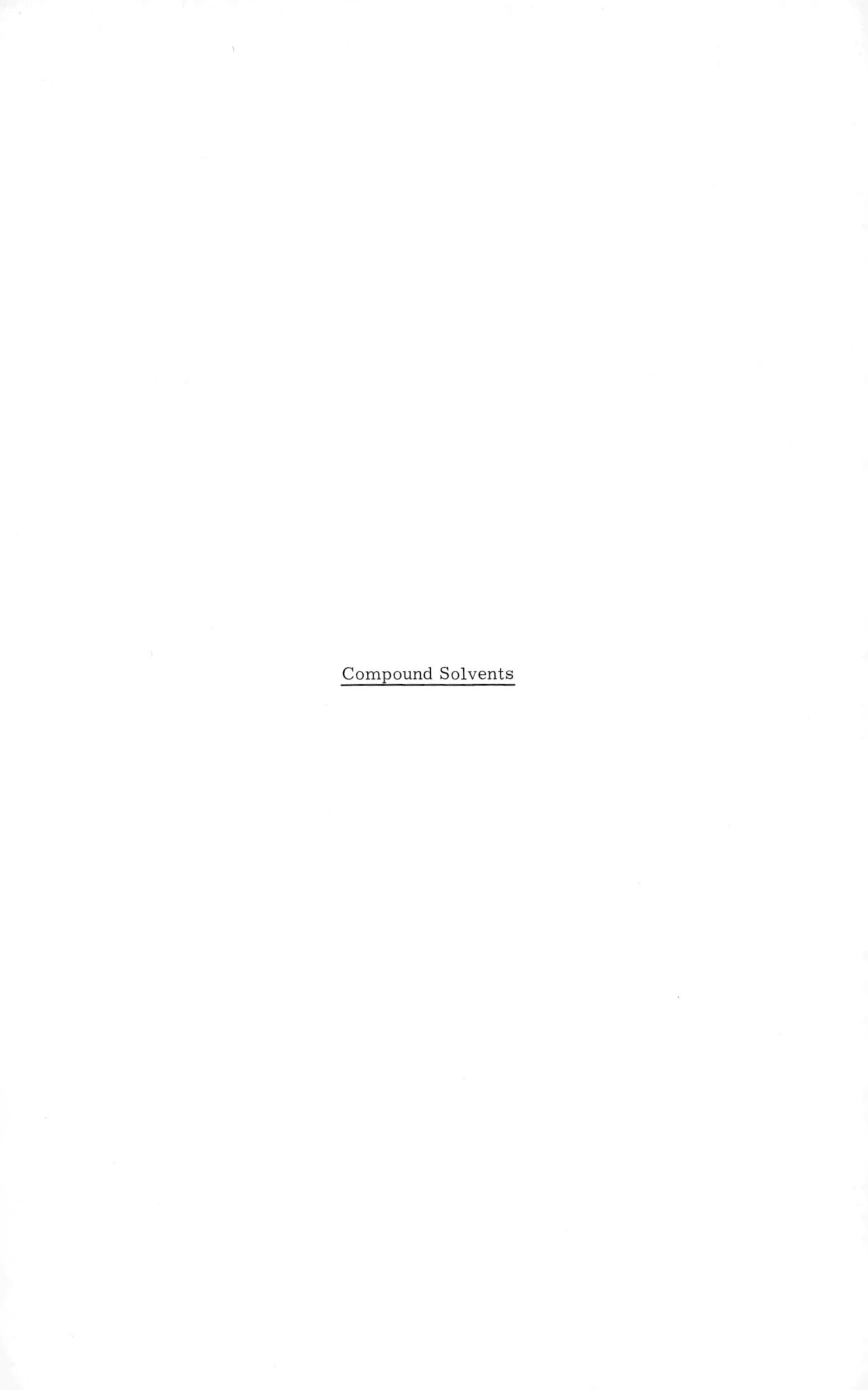

Compound Solvents

SOLVENT GaAs

Solute	Atomic % solute	Temp. (^{o}C.)	State	Solute accumulates at:	Transference no. (gm ion Faraday^{-1})	Electric mobility $cm^2 v^{-1} sec^{-1}$	Effective valence (Z^o)	Reference
Zn	trace	800 to 900	solid	C	-	at 860°C. 2.92×10^{-8}	-	85
Zn	trace	830 to 1130	"	C	-	approx. $(1 \text{ to } 10) \times 10^{-8}$	(+2)	140
Li	trace	740 to 980	"	C	-	$(1.6 \text{ to } 9.4) \times 10^{-5}$	+0.2 to +0.3	140
Cu	trace	820	"	C	-	0.98×10^{-5}	+0.6	140
Cu	trace	850	"	C	-	4.9×10^{-5}	-	140
Cu	trace	930	"	C	-	7.4×10^{-5}	+0.4	140
Cu	trace	950	"	C	-	9.9×10^{-5}	+0.4	140
Cu	trace	1000	"	C	-	8.5×10^{-5}	+0.3	140
Cu	heavily doped	830	"	C	-	-	+1.07	140
Cu	heavily doped	870	"	C	-	-	+0.94	140
Cd	trace	950	"	C	-	2.6×10^{-10}	-	140
Cd	trace	980	"	C	-	8×10^{-10}	-	140

SOLVENT InAs

Solute	Atomic % solute	Temp. ($^oC.$)	State	Solute accumulates at:	Transference no. (gm ion $Faraday^{-1}$)	Electric mobility $cm^2v^{-1}sec^{-1}$	Effective valence (Z^o)	Reference
Zn	trace	605	solid	C	-	0.326×10^{-6}	+2.06	130
Zn	ie(4.5 to 4.8)	665	"	C	-	0.68×10^{-6}	+2.0	130
Zn	$\times 10^{19}\ cm^{-3}$	698	"	C	-	0.733×10^{-6}	+1.6	130
Zn		735	"	C	-	1.32×10^{-6}	+1.85	130
Zn		757	"	C	-	1.45×10^{-6}	+1.8	130

SOLVENT Bi_2Te_3

Solute	Atomic % solute	Temp. ($^oC.$)	State	Solute accumulates at:	Transference no. (gm ion $Faraday^{-1}$)	Electric mobility $cm^2v^{-1}sec^{-1}$	Effective valence (Z^o)	Reference
Cu	trace	0 to 400	solid	A, C(a)	-	-	-	123
Ag	trace	0 to 400	"	A, C(a)	-	-	-	123
Au	trace	0 to 400	"	A, C(a)	-	-	-	123

(a) At C.D. of 150 A/cm^2, migration always towards C
At C.D. of 250 A/cm^2, migration towards C below 300^oC, to both A and C up to 400^oC, and towards A above 400^oC.

Section C

Isotope Separation in Pure Metals by Electrotransport. (Haeffner Effect)

GROUP	METAL	STATE	REFERENCE
(I)	Li	liquid	12,150,153
	K	"	12,148,151,152,153,159
	Rb	"	152,155,157,159
(II)	Zn	liquid	12,151,153
	Cd	"	12,151,153
	Hg	"	12,143,144,145,153
(III)	Ga	liquid	12,146,147,152,153,158
	In	"	12,149,153,154,158,159
	U	solid	144
(IV)	Sn	liquid	12,151,153,156,158
	Pb	"	156,158

In all cases the HEAVY ISOTOPE has been found to accumulate at the CATHODE

Section D

Electrotransport of Solid Intermetallic Compounds in Liquid Metals

LIQUID METAL	SOLID COMPOUND	TEMPERATURE (^{o}C)	REFERENCE
Cd	$Cd_{11}Ce$	460 (3 to 5 wt. % Ce)	161
In	In_3Ce	460 (3 to 5 wt. % Ce)	161
Sn	Sn_3Ce	410 (3 to 5 wt. % Ce)	161

The electrotransport force was found to be much less than the force of gravity and no separation was effected.

REFERENCES FOR ELECTROTRANSPORT DATA TABLE

1. K.E. Schwarz, "Elektrolytische Wanderung in flüssigen und festen Metallen", 1940, Leipzig, (J.A. Barth).

2. P.C. Mangelsdorf, "Electrolysis and Diffusion in Liquid Alloys", Met. Soc. A.I.M.E. Conferences, 7, 429; 1961 New York and London, (Interscience Publishers); J. of Chem. Physics, 1959, 30, 1170; and ibid., 1960, 33, 1151.

3. S.I. Drakin, T.N. Sergeeva, and U.N. Rusakova, Zhur. Fiz. Khim., 1961, 35, 551.

4. S.I. Drakin and A.K. Maltsev, Zhur. Fiz. Khim., 1957, 31, 2036.

5. J.C. Angus and E.E. Hucke, J. Phys. Chem. 1961, 65, 1549.

6. S.I. Drakin, Yu. K. Golubkova, and E.P. Ushakova, Zhur. Fiz. Khim., 1960, 34, 866; Russian J. Phys. Chem., 1960, 34, 411.

7. D.K.Belashchenko and G.A. Grigor'ev, Izvest. Vysshikh. Uchebn. Zaved. Chernaya Met., 1961, (11), 116; 1962, (5), 120.

8. D.K.Belashchenko, Izvest. Akad. Nauk. S.S.S.R. (Otdel Tekhu) Met. i Topl., 1960, (6), 89.

9. D.K.Belashchenko, Russian J. Phys. Chem., 1961, 35, 923.

10. M.A. Rabkin, J. Appl. Chem. U.S.S.R., 1957, 30, 832.

11. V.I. Yavoisky and G.I. Batalin, Trudy Nauchn. Tekhn. Abshchestva Chernoi Met., 1955, 4, 74.

12. J.D.Verhoeven,Met. Rev., 1963, 8, (31), 311.

13. V.A. Rotin, D.K.Belashchenko, B.S. Bokshtein, and A.A. Zhukovitsky, Zavod. Lab. 1964, 30, 186.

14. P.P. Kuz'menko, E.I. Khar'kov, and V.I. Lozovoi, Ukrain. Fiz. Zhur., 1964, 9, 881.

15. L.F. Ostrovsky and P.P. Kuz'menko, Physics of Metals and Metallography 1964, 17, 72.

16. T.N. Sergeeva and S.I. Drakin, Trudy Moskov. Khim. Teckhnd. Inst., 1962, (38), 103.

17. S.I. Drakin, A.M. Borisova, and W.M. Pugscewicz, Russian J. Phys. Chem., 1963, 37, 3.

18. S.I. Drakin, T.N. Sergeeva and A.I. Trepakov, Russian J. Phys. Chem., 1964, 38, 170.

19. C. Wagner and G. Heller, Z. Phys. Chem. (B), 1940, 46, 242.

20. D.R. Hay and E. Scala, Trans. Met. Soc. A.I.M.E., 1965, 233, 1153.

21. J.M. Williams and C.L. Huffine, Nuc. Sci. Eng., 1961, 9, 500.

22. Th. Heumann, "The Physical Chemistry of Metallic Solutions and Intermetallic Compounds". N.P.L. Symposium, 1959, Paper 2C, H.M.S.O. London.

23. P.S. Ho and H.B. Huntington, J. Phys. Chem. Solids, 1966, 27, 1319.

24. R.P. Johnson, Phys. Rev., 1938, 53, 766 and 1938, 54, 459.

25. J.D. Verhoeven and E.E. Hucke, Trans. Quarterly A.S.M., 1962, 55 866.

26. J.D. Verhoeven and E.E. Hucke, Trans. Met. Soc. A.I.M.E., 1963, 227, 1156.

27. W. Seith and H. Wever, Z. Elektrochem., 1953, 57, 891.

28. C.S. Fuller and J.C. Severiens, Phys. Rev., 1954, 96, 21.

29. O. Kubaschewski and K. Reinartz, Z. Elektrochem., 1948, 52, 75.

30. W. Seith and O. Kubaschewski, Z. Elektrochem., 1935, 41, 551.

31. P. Dayal and L.S. Darken, J. Met., 1950, 188, 1156.

32. W. Seith and H. Etzold, Z. Elektrochem., 1934, 40, 828; 1935, 41, 122.

33. K.E. Schwarz and R. Stockert, Z. Elektrochem., 1939, 45, 464.

34. I.H. De Boer and J.D. Fast, Rec. trav. chim., 1940, 59, 161.

35. F. Claisse and H.P. König, Acta Met., 1956, 4, 650.

36. H. Wever, Deutsche Gesellschaft für Metallkunde, Sept. 1957.

37. B. Baronowski, Bulletin de l'Acad. Polonaise Sci., 1956, IV, 456.

38. F.P. Golotyuk, R.P. Kuz'menko and E.I. Khar'kov, Physics of Metals and Metallography, 1965, 19, 78.

39. S.G. Epstein, Trans. A.I.M.E., 1966, 236, 1123.

40. W. Jost, "Diffusion in Solids, Liquids and Gases", 1960, New York, (Academic Press).

41. I.N. Frantsevich, D.F. Kalinovich, I.I. Kovensky and M.D. Smolin, Soviet Physics - Solid State, 1959., 1, 58.

42. S.D. Gertsriken, I. Ya. Dekhtyar, V.S. Mikhalenkov and V.M. Fal'chenko Ukrain.Fiz. Zhur., 1961, 6, 129.

43. S.D. Gertsriken, I. Ya. Dekhtyar, V.S. Mikhalenkov and E.G. Madatova, ibid., 1960, 5, 79.

44. D.F. Kalinovich, Soviet Physics - Solid State, 1961, 3, 812.

45. D.F. Kalinovich, I.I. Kovensky, M.D. Smolin and I.N. Frantsevich, Physics, Metals and Metallography, 1960, 10, (1), 41.

46. D.F. Kalinovich, I.I. Kovensky, M.D. Smolin and I.N. Frantsevich, Inzhen. Fiz. Zhur., Akad. Nauk. Belorussk, S.S.S.R., 1961, 4, (5), 108.

47. I.N. Frantsevich, D.F. Kalinovich. I.I. Kovensky and M.D. Smolin, ibid., 1959, 2, (4), 47.

48. D.F. Kalinovich, I.I. Kovensky and M.D. Smolin, Physics Metals and Metallography, 1961, 11, (2), 148.

49. I.N. Frantsevich, "An investigation of Electric Transport in Solid Metals using Radioactive Isotopes", 1957, London (Pergamon Press).

50. I.N. Frantsevich and I.I. Kovensky, Dopovidi Akad, Nauk. Ukr. R.S.R., 1961, (9), 1169.

51. I.N. Frantsevich, D.F. Kalinovich, I.I. Kovensky and M.D. Smolin, Izvest. Akad. Nauk S.S.S.R. (Otdel. Tekhn.)Met. i Topl. 1959, (1), 71.

52. I.N. Frantsevich, D.F. Kalinovich, I.I. Kovensky, V.V. Pen'kovesky and M.D. Smolin, Dopovidi Akad. Nauk Ukr. R.S.R., 1958, (7), 736.

53. I.N. Frantsevich, D.F. Kalinovich, I.I. Kovensky and M.D. Smolin, Inzhen. Fiz. Zhur., Akad. Nauk Belorussk. S.S.S.R., 1959, 2, (9), 62.

54. I.N. Frantsevich, D.F. Kalinovich, I.I. Kovensky and V.V. Pen'kovsky, Ukrain. Fiz. Zhur., 1958, 3, (1), Suppl. 64.

55. D.F. Kalinovich, I.I. Kovensky and M.D. Smolin, Physics of Metals and Metallography, 1962, 13, 120.

56. I.N. Frantsevich and I.I. Kovensky, Dopovidi Akad. Nauk Ukr. R.S.R., 1961, (11), 1471.

57. M.D. Smolin and I.N. Frantsevich, Soviet Physics - Solid State, 1962, 3, 1536.

58. M.D. Smolin and I.N. Frantsevich, Soviet Physics - Doklady, 1961, 6, 66.

59. I.N. Frantsevich and M.D. Smolin, Dopovidi Akad. Nauk Ukr. R.S.R., 1961, (7), 908.

60. P.P. Kuz'menko, L.F. Ostrovsky and V.S. Koval'chuk, Physics of Metals and Metallography, 1962, 13, 83.

61. E.I. Khar'kov and P.P. Kuz'menko, Ukrain. Fiz. Zhur., 1959, 4, 389.

62. P. P. Kuzmenko, E. I. Kharkov and G. P. Grinevich, ibid., 1960, 5, 683.

63. P. P. Kuzmenko, and L. F. Ostrovsky, ibid., 1961, 6, 525

64. P. P. Kuzmenko, L. F. Ostrovsky and V. S. Kovalchuk, Soviet Physics-Solid State, 1962, 4, 356

65. C. J.Gallagher, J. Phys. Chem. Solids, 1957, 3, 82.

66. I. I. Ibragov and A. A. Kuliev, Soviet Physics - Solid State, 1962, 3, 2418

67. B. I. Boltaks, G. S. Kulikov and R. Sh. Malkovich, ibid., 1960, 2, 2134

68. B. P. Konstantinov and L. A. Badenko, ibid., 1961, 2, 2400

69. D. K. Belashchenko and G. A. Grigorev, Izvest, Vysshikh Uchebn. Zaved. Chernaya Met., 1962, (1), 124

70. G. N. Lewis, E. O. Adams and E. H. Lanman, J. American Chem. Soc., 1915, 37, 2656

71. K. P. Romadin, "Electromigration in Liquid and Solid Metallic Solutions". No. 167, Trudy V. V. I. A. im Zhukovskogo, Moscow, 1947.

72. B. Kremann, A. Vogrin and H. Scheibel, Monatsh. für Chemie, 1931, 57, 323

73. B. Kremann, F. Bauer, A. Vogrin and H. Scheibel, ibid., 1930, 56, 35

74. I. N. Frantsevich, D. F. Kalinovich, I. I. Kovensky and M. D. Smolin, Soviet Physics - Solid State, 1963, 5, (5), 905

75. D. K.Belashchenko, Russian Chemical Reviews, 1965, 34, 219

76. G. A. Grigorev and D. K. Belashchenko, Izvest. Vysshikh. Uchebn. Zaved. Chernaya Met., 1962, 137

77. D. K. Belashchenko, and C. Jung-sheng, Russian J. Phys. Chem., 1963, 37 593

78. G. Nehlep, W. Jost and R. Linke, Z. Elektrochem., 1936, 42, 150

79. W. Seith, Diffusion in Metallen, (Platzwechselreaktionen) 2nd Ed., 1955 Berlin (Springer Verlag).

80. A. Cohen and W. Specht, Z. Physik 1930, 62, 1.

81. W. Jost and R. Linke Z. Physik. Chem., 1935, B29, 127

82. W. Seith and Th. Daur, Z. Elektrochem., 1938, 44, 256

83. J. Angus, J. Verhoeven and E. E. Hucke, Trans. A. I. M. E., 1959, 7, 447

84. Yu. F. Babikova, P. L. Gruzin et al., Fiz. Metall. i Metalloved., 1957 5, 255

85. V. A. Dhaka "Electromigration of Impurities in Si and GaAs", Electrochem. Soc. Conference: Impurity Diffusion in Semiconductors" New York, 1963

86. R.S. Malkovich and N.A. Alimbarashvili, Soviet Physics - Solid State, 1963, 4, 1725.

87. N.I. Ibragimov, M.G. Shakhtakhtinskii and A.A. Kuliev, ibid., 1963, 4, 2430.

88. H.B. Huntington and A.R. Grone, J. Phys. Chem. Solids, 1961, 20, 76.

89. H. Wever and W. Seith, Zeit. Elektrochem. 1955, 59, 942.

90. H. Wever, ibid., 1956, 60, 1170.

91. H. Wever, "The Physical Chemistry of Metallic Solutions and Intermetallic Compounds". Vol. 1, paper 2L, 1959, London. (H.M.S.O.).

92. P.P. Kuz'menko and E.I. Khar'kov, Ukrain. Fiz. Zhur., 1958, 3, 528; 1959, 4, 401; 1959, 4, 537; 1960, 5, 428; 1961, 6, 140; 1962, 7, 117.

93. M. Gerardin, Compt. Rend., 1861, 53, 727.

94. S.G. Epstein and A. Paskin, Physics Letters, 1967, 24A, 309.

95. S.G. Epstein, "The properties of Liquid Metals", Proceedings of International Conference, 1966, Brookhaven Nat. Lab., New York, Taylor + Francis Limited, London, 1967; (Also Advances in Physics, 1967, 16).

96. S.I. Drakin and Yu. K. Titova, Russian J. Phys. Chem., 1967, 41, 319.

97. S.G. Epstein, Trans. A.I.M.E., 1967, 239, 627.

98. D.T.J. Hurle, J.B. Mullin and E.R. Pike, Phil. Mag., 1964, 9, 423.

99. D.T.J. Hurle, J.B. Mullin and E.R. Pike, J. of Materials Science, 1967, 2, 46.

100. F.Y. Lieu, Acta Met., 1967, 15, 1405.

101. V.A. Mikhaylov, R.A. Polovinkina, S.I. Drakin and G.M. Frolova, Physics of Metals and Metallography, 1966, 22, 63.

102. A. Lodding, J. Phys. Chem. Solids, 1965, 26, 143.

103. H. Hering and H. Wever, Acta Met., 1967, 15, 377.

104. Th. Hehenkamp, Acta Met., 1966, 14, 887.

105. O.N. Carlson, F.A. Schmidt and D.T. Peterson, J. Less Common Metals 1966, 10, 1.

106. R.S. Wagner, C.E. Miller and H. Brown, Trans. A.I.M.E., 1966, 236, 554.

107. G.M. Neumann and W. Hirschwold, Z. Naturforsch, 1967, 22A, 388.

108. W. Kleinn, Dissertation, Technische Hochschule Hannover, 1958.

109. Th. Hehenkamp, Ch. Herzig and Th. Heumann, Symposium on Electrotransport and Thermodiffusion in Metals, Münster, 1965.

110. D. Lazarus and H.M. Gilder, Phys. Rev., 1966, 145, 507.

111. H. Dübler and H. Wever, Physica Status Solidi, 1968, 25, 109.

112. H.B. Huntington, J. Phys. Soc. Japan, 1963, 18, Supplement 11, 202.

113. H. Hering and H. Wever, Tagung bayer. Phys. Ges. Munich, 1964.

114. P.S. Ho, J. Phys. Chem. Solids, 1966, 27, 1331.

115. R. V. Penney, J. Phys. Chem. Solids, 1964, 25, 335.

116. A.G. Grone, J. Phys. Chem. Solids, 1961, 20, 88.

117. G.A. Sullivan, J. Phys. Chem. Solids, 1967, 28, 347.

118. R.A. Oriani and O.D. Gonzalez, Trans. A.I.M.E., 1967, 239, 1041.

119. G.A. Sullivan, Phys. Rev., 1967, 154, 605.

120. G. Lormand, J.C. Rouais, M. Lallemand and C. Eyraud, Revue Metallurgie, 1967, 64, 59.

121. H. Hering and H. Wever, Z. Physik Chemie., 1967, 53, 310.

122. R.H. Moore, F.M. Smith and J.R. Morrey, Trans. A.I.M.E., 1965, 233 1259.

123. H.P. Dibbs and J.D. Keys, Canad. J. Physics, 1967, 45, 3945.

124. M.D. Smolin, Soviet Physics - Solid State, 1967, 9, 1415.

125. V.A. Sterkhov and P.V. Pavlov, Soviet Physics - Solid State, 1967, 9, 495.

126. V.A. Sterkhov, V.A. Panteleev and P.V. Pavlov, Soviet Physics - Solid State, 1967, 9, 533.

127. Th. Heumann and H. Meiners, Z. Metallkunke, 1966, 57, 571.

128. F.A. Schmidt and J.C. Warner, J. Less-common Metals, 1967, 13, 493.

129. D.T. Peterson, F.A. Schmidt and J.D. Verhoeven, Trans. A.I.M.E., 1966, 236, 1311.

130. B.I. Boltaks and S.I. Rembeza, Soviet Physics - Solid State, 1967, 8, 2117.

131. W. Dornelas and P. Lacombe, Compt. Rend., 1967, 265, 359.

132. S. Iida, Jap. J. Applied Physics, 1967, 6, 77.

133. R.L. Schmidt and J.D. Verhoeven, Trans. A.I.M.E., 1967, 239, 148.

134. H.J. Stepper and H. Wever, J. Phys. Chem. Solids, 1967, 28, 1103.

135. I.A. Blech and E.S. Meieran, Applied Physics Letters, 1967, 11, 264.

136. D. Chhabra, N. Ainslie and D. Jepsen, The Electrochemical Society Meeting, Dallas, Texas, May, 1967.

137. R.G. Shepheard and R.P. Sopher, ibid.

138. W.E. Mutter, ibid.

139. Th. Heumann and H. Stüer, Z. Naturforsch, 1967, 22A, 1184.

140. B.I. Boltaks and T.D. Dzhafavov, Physica Status Solidi, 1967, 19, 705.

141. M.J. Bibby and W.V. Youdelis, Canad. J. Physics, 1966, 44, 2363.

142. M.J. Bibby, L.C. Hutchinson and W.V. Youdelis, Canad. J. Phys., 1966, 44, 2375.

143. E. Haeffner, Nature, 1953, 172, 775.

144. E. Haeffner, Th. Sjoborg and S. Lindhe, Zeit. Naturforsch., 1956, 11A, 71.

145. I.V. Bogoyarlensky, V.N. Grigor'ev, N.S. Rudemko and D.G. Dolgopolov, Soviet Physics - JETP, 1958, 6, 450; 1959, 37, 1241; and 1960, 10, 884.

146. G. Nief and E. Roth, Compt. rend., 1954, 239, 162.

147. M.M. Goldman, G. Nief and E. Roth, ibid., 1956, 243, 1414.

148. A. Lunden, C. Reutersward and A. Lodding, Z. Naturforsch, 1955, 10A, 924.

149. A. Lodding, A. Lunden and H. von Ubisch, ibid., 1956, 11A, 139.

150. A. Lunden, A. Lodding and W. Fischer, ibid., 1957, 12, 268.

151. A. Lodding, ibid., 1957, 12A, 569; 1959, 14A, 7; 1959, 14A, 934; and 1961, 16A, 1252.

152. A. Lodding, J. Phys. Chem. Solids, 1967, 28, 557.

153. A. Lodding, Proceedings of the International Symposium on Isotope Separation, North Holland Publishing Co., Amsterdam, 1958, 308.

154. A. Lodding, Gothenburg Stud. Phys., 1961, 1.

155. A. Lodding, J. Chim. Phys. 1963, 60, 254.

156. P.P. Kuz'menko, Symposium on Electrotransport and Thermodiffusion in Metals, Munster, 1965.

157. A. Norden and A. Lodding, Z. Naturforsch, 1967, 22A, 215.

158. P.P. Kuz'menko, E.I. Khar'kov and V.I. Lozovoi, Doklady Akad. Nauk. S.S.S.R., 1965, 160, 1343.

159. A. Lodding and A. Klemm, Z. Naturforsch, 1962, 17A, 1085.

160. V.A. Mikhailov, M.V. Kornievich and R.A. Polovinkina, Doklady Akad. Nauk. S.S.S.R., 1966, 171, 147.

161. D.K. Belashchenko and I.A. Magidson, Physics of Metals and Metallography, 1966, 22, 144.

162. S.G. Epstein, Phys. and Chem. Liquids, 1968, 1, 109.

163. D.R. Packard and J.D. Verhoeven, Ames Laboratory, U.S. Atomic Energy Commission (Ames Iowa), Contribution No. 2218.

164. N.K. Archipova, S.M. Klotsman, A.N. Timofeev and I. Sh. Trakhtenberg Phys. Stat. Solidi, 1966, 16, 729.

165. M. Le Blanc and R. Jäckh, Zeit. für Elektrochem., 1929, 35, 395.

166. R.G.R. Sellors, Ph. D. Thesis, Birmingham University, 1968.

167. S.G. Epstein, Trans. A.I.M.E., 1968, 242, 1771.

168. B. Spriet, Commissariat a l'Energie Atomique, Bibliographie, No. 74, 1966.

169. Y.F. Babikova and P.L. Gruzin, Metallurgy and Metallography of Pure Metals, p. 125.

170. B. Spriet, J. Mat. Nucl., 1965, 15, 220.

171. D.T. Peterson and F.A. Schmidt, J. Metals, 1965, 17, 107.

172. J.D. Marchant, E.S. Shedd and T.A. Henrie, Proc. 2nd Rare Earth Research Conference, p. 143, (Gordon + Breach), New York, 1962.

173. R.E. Hummel and H.M. Breitling, Z. Naturforsch. 1971, 26A, 36.

174. P.S. Ho and L.D. Glowinski, ibid., 1971, 26A, 32.

175. W.B. Alexander, ibid., 1971, 26A, 18.

176. T. Okabe and A.G. Guy, Met. Trans., 1970, 1, 2705

177. Y. Adda and J. Philibert, "La Diffusion dans les Solides", Tome II, p. 919, Presses Universitaires de France, Paris, 1966.

178. D.H. Laananen and W.B. Nowak, Rev. Sci. Inst., 1969, 40, 422.

179. N. Van Doan and G. Brebec, J. Phys. Chem. Solids, 1970, 31, 475.

180. I.I. Kovenskii, Soviet Physics - Solid State, 1961, 3, 812.

181. Yu. K. Titova, S.I. Drakin and I.B. Alikina, Russian J. Phys. Chem., 1968, 42, 1194.

182. D.T. Peterson and F.A. Schmidt, J. Less Common Metals, 1969, 18, 111.

183. D.N. Vasilovskii and I.V. Zakurdaev, Fiz. Tverd Tela, 1968, 10, 3640.

184. J.C. Jousset and H.B. Huntington, Physica Status Solidi, 1969, 31, 775.

185. I.N. Larionov, N.M. Roizin, V.M. Nogin and E.T. Avrasin, Soviet Physics - Semiconductors, 1968, 1, 1175.

186. A.V. Vanyukov, D.K.Belashchenko and U.K. Samedinov, Fiz. Metal. Metalloved., 1970, 29, 182.

187. D.F. Kalinovich, I.I. Kovenskii, M.D. Smolin and V.M. Statsenko, Soviet Physics - Solid State, 1969, 11, 1397 and Fiz. Metal. Metalloved. 1970, 29, 653.

188. D.F. Kalinovich, I.I. Kovenskii and M.D. Smolin, Fiz. Metal. Metallo-ved., 1968, 26, 762, Physics of Metals and Metallog., 1968, 26 (4), 192.

189. P.P. Kuz'menko, E.I. Khar'kov, L.M. Korochkina and T.D. Kotikova, Fiz. Metal. Metalloved., 1967, 23, 556.

190. S.G. Epstein and J.M. Dickey, Phys. Rev. B. 1970, 1, 2442.

191. J.C. Peacock and A.D, Wilson, J. Appl. Physics, 1968, 39, 6037.

192. I. A. Blech and E.S. Meieran, J. Appl. Physics, 1969, 40, 485.

193. G. Guy, Trans. A.I.M.E., 1969, 245, 2221.

194. Ch. Herzig, Thesis (Diplomarbeit), University of Münster, Germany, Oct. 1965.

195. A. Herold and J.C. Rat, Compt. Rend. (C), 1970, 271, 701.

196. D.F. Kalinovich, I.I. Kovenskii, M.D. Smolin and V.M. Statsenko, Fiz. Tverd. Tela., 1969, 11, 2378, Soviet Phys. - Solid State, 1970, 11, 1919.

197. D.F. Kalinovich, I.I. Kovenskii and M.D. Smolin, Fiz. Tverd. Tela., 1970, 12, 929; Soviet Phys. - Solid State, 1970, 12, 723; Fiz. Met. Metalloved., 1970, 29, 671.

198. T.S. Lakshmanan, D.L. Olson and D.A. Rigney, Scripta Met., 1971, 5 1099.

199. J.L. Blough, D.L. Olson and D.A. Rigney, J. App. Phys. 1972, 43, 2476.

200. R. Kremann and H. Scheibel, Monatsch. für Chemie, 1931, 57, 241.

201. L. Lämmermayr, Monatsch. für Chemie, 1933, 62, 67.

202. R. Kremann and W. Piwetz, Monatsch. für Chemie, 1929, 53/54, 203.

203. R. Kremann and E.I. Schwarz, Monatsch. für Chemie, 1930, 56, 26.

204. R. Kremann and W. Piwetz, Monatsch. für Chemie, 1930, 56, 71.

205. R. Kremann and H. Lämmermayr, Monatsch. für Chemie, 1933, 62, 61.

206. N. Van Doan, J. Phys. Chem. Solids, 1971, 32, 2135.

207. N. Van Doan, J. Phys. Chem. Solids, 1970, 31, 2079.

208. J.L. Routbort, Phys. Rev., 1968, 176, 796.

209. W.B. Alexander, Proc. Europhys. Conf. "Atomic Transport in Solids and Liquids", Marstrand, 1970, p. 97 (Verlag. Z. für Naturforsch., Tubingen, 1971).

210. J.F. D'Amico and H.B. Huntington, J. Phys. Chem. Solids, 1969, 30, 2607.

211. D.R. Campbell and H.B. Huntington, Phys. Rev., 1969, 179, 601.

212. J.C. Pieri, J. Bagnol and E. Berger, Comptes Rend., Serie C, 1971, 273, 946.

213. R. Kremann, B. Korth and E.I. Schwarz, Monatsch für Chemie, 1930, 56, 16.

214. K. Dreyer, C. Herzig and T. Heumann, Proc. Europhysics Conf. "Atomic Transport in Solids and Liquids", Marstrand, 1970, p. 61 (Verlag. Z. für Naturforsch., Tubingen, 1971).

215. G.L. Hofman and A.G. Guy, J. Phys. Chem. Solids, 1972, 33, 2167.

216. D. Grimme, Proc. Europhysics Conf. "Atomic Transport in Solids and Liquids", Marstrand, 1970, p. 65.

217. T. Hehenkamp, ibid., 1970, p. 68.

218. I.N. Frantsevich, D.F. Kalinovich, I. Kovenskii and M.D. Smolin, ibid. 1970, p. 100.

219. D.K.Belashchenko, ibid., 1970, p. 173.

220. P. Thernquist and A. Lodding, Z. für Naturforsch., 1968, 23a, 627.

221. I.A. Magidson, Izv. Vyssh. Ucheb. Zaved. Tsvet. Met., 1967, 10, 75.

222. H.R. Patil and H.B. Huntington, J. Phys. Chem. Solids, 1970, 31, 463.

223. F.M. D'Heurle, Met. Trans., 1971, 2, 683.

SURFACE SELF - DIFFUSION OF METALS

G. Neumann and G. M. Neumann 1972

International Standard Book Number 0-87849-501-0

Library of Congress Catalog No. 72-075071

132 Pages, 210 References, 35 Figures, Tables with the available diffusion data measured from 1949 to 1971. Paperback – US $ 17.50.

The book represents a highly authoritative and complete account of our present knowledge of atomic transport on metal surfaces. Covered are

Experimental Techniques: Sintering experiments; field-electron-emission-microscopy; mass transfer techniques; radioactive tracer methods; field-ion-emission-microscopy; techniques of evaluation for the various methods.

Experimental Data: Listing of all available data; orientation dependence of the surface diffusion coefficient; effect of impurity adsorption on the surface diffusion mobility; phenomenological relations for the activation energy of surface diffusion.

Metal Surface Structure: Kossel-Stranski model of crystal surfaces; Terrace-Ledge-Kink model of surface diffusion; interaction potential models.

Mechanisms of Surface Diffusion: Analysis of available data and theories with respect to transport mechanisms.

The present volume is part of the DIFFUSION MONOGRAPH SERIES described on the inside cover.